COME OUT FIGHTING

COME OUT FIGHTING

Combat Infantry in World War II
as told by Staff Sergeant
HENRY C.STRECKER

Deeds Publishing | Athens

Published by Deeds Publishing in Athens, GA
www.deedspublishing.com

Printed in The United States of America

Cover design by Mark Babcock.

ISBN 978-1-961505-22-3

Books are available in quantity for promotional or premium use. For information, email info@deedspublishing.com.

First Edition, 2024

10 9 8 7 6 5 4 3 2 1

To our parents Leanora and Henry Strecker

Shaped by our parents, Leanora "Dee" and Henry Strecker, in the wake of World War II, we learned from our mother the wonders of nature around us, the power of books, poetry, and art — and from our father's voice, the love in a song and his courage in a prayer. This started for us at a very young age in our one-bedroom apartment. It was a small space, but filled with love for each other, and their two children Leslie and Steve.

We met the outside world on our frequent walks to Arlington Memorial Gardens just a few blocks from home. The trail and pond landscape were filled with the promise of a curiosity for nature that came to life at Mom's direction — "Oh look at this leaf … see the frog … here's a bug … a rock … a flower … " How adventurous our little world became. Mom read us the poem by Joyce Kilmer "Trees" which was etched on a sculpture and the trees towered above us right there at Arlington and the wider world, like the poppy fields of Flanders, were all part of our life ever after.

Dad taught us about the stars and birds, how to row a boat, how to help do oil changes on cars, to "Be Prepared" and so much more. Our dinner table was a place for conversations about history, the news of the day, always learning, always sharing. Henry sang us to sleep at night with "Froggy went a Courtin", the Tennessee Waltz, and Lilli Marlene. He led us through the harrowing days of homework. He will always be that young man of 1944 in my mind, overseas in a surreal world before our lives began. When I was in fifth grade, he would share some of his experiences from the war. In November, he would often be in tears reliving his memories. Like so many veterans from the Civil War onward, our homes were affected by very real trauma.

Despite the psychological stress he suffered, like so many men of his generation, he came through in everyday life for his family.

Love you always and forever Dad.

Prelude

They were somewhere in France. Someone had left the field telephone down in a forested valley at dusk. And the someone in charge wanted it brought back out. He told Sergeant Strecker to go get the phone. German snipers were whistling to one another like birds in the darkness.

Hank said, "No, I won't do that tonight. I will go in the morning to get it."

So, the man in charge said, "Do you know what your job is in this outfit, Strecker?"

To which Hank replied, "Yes sir, I take over if you get killed."

Henry Carl Strecker was beginning his sophomore year at Mt. Healthy High School, near Cincinnati, Ohio. His days of youthful scouting and hunting and growing up in the farm country around Finneytown were almost over. It was early September.

As he slipped through the upstairs bedroom, he overheard his Mom and Dad quietly, resignedly, discussing September 1, 1939 — the German invasion of Poland. News of it had just come over the radio. The threat of war had become reality. At the time, Hank wasn't too upset by the news. He could not foresee the profound effect it would have on his life and the lives of millions — most just kids like himself. And so, his life at 1038 North Bend Rd. went on as usual.

Henry, and later his brother Edward, bicycled the rent at twenty-two dollars a month, to the new owners (the Longman's) who lived on Corcoran Place. The Strecker's lived in the house over twenty years.

The Federal style brick farm house with its five fireplaces, dated from about 1802. The fireplaces included one in the basement, set waist high on the wall with a surrounding work area. The largest one was in the dining room — a huge walk-in type for large kettles. The land had been a horticultural paradise, full of flowers, berries, elderberry, gooseberry, blackberry, and many fruit trees … apple and pear. A huge pear tree towered over the side door. There was also a fresh water spring that bubbled up behind the house on the seven acres. It was invaluable when the city water supply from the Ohio River and Mill Creek valley became contaminated by flooding in 1937.

No one could recall the original homeowners' name, but they were surely dedicated to this horticultural paradise. The farmhouse was later restored by Paul Bartels and became the Baroque Violin Shop at 1038 North Bend Road near Daly Road. The home now has a historic designation, thanks to Mr. Bartels hard work. The seven acres behind the house evolved into a street of bungalows where the basements are noted for being wet, thanks to the spring water.

(There was one other residence also noted for its horticulture, several miles away which was owned by Rabbi Isaac Wise in 1861, forty acres with many crops, livestock, and orchards. He became the founder of Hebrew Union College in Cincinnati.)

Henry Strecker was born on June 19, 1922 in Cincinnati, Ohio, the eldest son of Lillie Dreier and Henry Strecker, who started work as a lithographer for Frank Tuchfarber and then went on to become a hard-working machinist. They had lived in a small apartment on Elgin Place in Mt. Adams, moved to Beekman St,. near the German part of town known as "Over the Rhine" and then moved to the upper floor of the spacious farmhouse in Finneytown where Grandma Elizabeth Sieckmeier and Grandpa Gustav Strecker, both German immigrants lived.

Gustav returned to Germany for family reunions twice, the last time in 1913, one year before World War I started. His brother Heinrich had lived in the U.S. for several years and returned to Germany and the family had many other cousins and relatives all over Germany. Their home area was Heiligenstadt and Dinglestadt in the Eichsfeld area of Thuringen. For years, the Streckers held very large family reunions, thanks to the work of Dr. Karl Strecker who traced their genealogy back to 1479. The last was in 1939.

Gustav and Elizabeth's children were Henry Sr., Otto, Walter, Hilda, Gustav Jr., Carl, Elsie, and Marie. Gustav Jr. who had Down's syndrome, died at the age of sixteen, when the pharmacy sent him an adult dose of codeine for a bout of pneumonia. Elizabeth never got over losing him.

In 1930, they still had no radio, so the boys enjoyed playing musical instruments together and with the German Donauschwaben Club which still exists today. Henry Sr. played English concertina and his brother Walter picked up the mandolin and guitar and are seen side by side in many old photos.

Henry, with his sisters Lillian and Ellen and his brothers, Alan and Edward grew up in a world surrounded by fields and woods with hunting and fishing and working hard at home. As they got older, they worked at places like the nearby farmer's fields, the Haskins garage next door, and the Rahn greenhouse nursery, which is still in business today.

Grandpa Gustav worked as an interpreter for the German immigrants in the court system in Cincinnati. He also worked at the Pullman factory as a young man and later as a private tutor. He had trained as a brewer in Germany as a young man.

Gustav's father, Hugo, was a doctor in Worbis and later Dingelstadt, who often helped railroad men for free in the days before insurance. He had studied at Wurzburg, Bonn, and Halle for his medical license. Hugo's grandfather, Philip, had run away from home with a best friend, the night before he was scheduled by his father, Christian, to go to the local monastery at Heiligenstadt, for a spiritual life. Phillip joined a French hussar cavalry unit with his friend during the Seven Years War, was shipped to the Netherlands, served for ten years, and finally came home to start up an overnight freight wagon hostel—a truck stop—Zum Adler (The Eagle). Often, there were as many as thirty wagons parked out front. Later he became a district administrator. His grandson, the doctor Hugo Strecker, did a successful cataract surgery on the old cavalryman who could still get on a horse at age eighty.

His great-grandson, Gustav in Finneytown, experimented with a recipe he concocted for shoe polish and later sold it for $350.00. One year one of his milk cows had a calf in the fall and she died. He carried the calf into the basement to raise through the winter and later lifted it outside again in the spring.

Their neighbors, the Erhardt family had a beautiful brick home down the road with a dairy farm. Their two boys, Fred and Dave, were close friends with Henry and their father often slept in the dairy barn with his prized Holsteins. One year several tested positive for milk fever and the Health Dept. made him destroy his beloved dairy herd. Mr. Erhardt broke down crying. The dairy started over and his son Fred later took some of the remaining cows out to his own farm far north of the city when he moved to Somerville, Ohio.

In summer there was baseball, where Henry covered "the hot corner" for the Mt. Healthy Owls. In 1941, according to the Zem-Zem yearbook, "In the first half of the eighth inning, Henry Strecker, smashed a home run into right center field to win the game for the Mounts." In autumn, there was hunting and in winter, ice skating. He loved to make model airplanes, studied planes, and eagerly watched planes fly overhead. Once in a while, they landed down the road in the open fields near Daly Rd.

Henry and Edward joined the Boy Scout Troop #390 headed by Ernie Staubach. They learned survival skills and had many good times at meetings and on camping trips as far away as Indian Lake in Ohio. Dad's sisters bought him a Scout uniform with hat and even a treasured and expensive bugle. He bought a pair of black leather gloves from Montgomery Ward for $12.00 and kept them all his life — just like the bugle — in perfect condition.

Scouting took up much of Henry's time and proved invaluable in keeping him physically and mentally sound in combat. He never forgot the influence and training of his scoutmaster, Ernie Staubach.

Another local neighbor had spent his youth as a cowboy out West. A not romantic life. He returned to his hometown for retirement and lived without electricity, relying on his fireplace, and slept on the floor, using his saddle as a pillow near the fire. The next generation would change the country forever — and very swiftly.

In 1940, on a stormy rainy day, their much loved, mother, Lillie Dreier died during a routine goiter operation. The children were told someone had accidentally caused her death during the operation and they never got over the shock. Even as adults, they wept over her death. Lillian stepped in to become mother to all the family. She had a unique gift of love and patience with her siblings and all the nieces and nephews who followed. They were very close-knit. The "Sisters" Lillian and Ellen lived together all their lives.

By 1941, the German armed forces had already built Fortress Europe. The conquered could only hope that the Americans would come. The tired, toughened, down-to-mud dogfaces who finally did come were more than just

American G.I.'s to the victims of Hitler's New Order. They were the Liberators.

Henry C. Strecker would be one of the Liberators. That same year, a stunned and angry America was swept into the war. Hank knew that he would be called up after his graduation in 1942. He smiled from the graduation photos, wearing a new canary yellow blazer, alongside his two brothers. All three of them later became sergeants in the army.

The war was not a popular idea with the American people in 1941, despite the total commitment they mustered. They had fought in World War I as the "war to end all wars" and the depression had tightened their belts. After Henry Strecker Sr. had spent all his savings ($4,000.00) to survive during the depression, the children had to get welfare for food. A truck would pull up to deliver milk, cheese, and peanut butter. It helped supplement the home-grown food and rabbit meat they raised. Milk from their two cows was kept outside on the window sill to keep it cool. Hot water was created by heating water on the wood fired stove. The windows frosted over from cold in the winter.

When war was declared, Henry went downtown to the Marine Corp center with a buddy, Carl Rahn, and applied. Carl passed the strict physical requirements, but the quota was full, and he would not be needed for several weeks. By that time the army would enlist him. (Carl's family still owns and works their thriving nursery near Spring Grove Cemetery.) Henry, with a weak left eye, failed to meet the

Marine standard, so the army would get him too. Next, he went to a sparse but trim army office on 4th St. for a physical. With a group of future GIs, he was examined and then told he would have to go to Fort Thomas for a blood test.

After several boring hours of waiting, they rode across the Ohio river in a rickety army bus to the sprawling grounds and imposing red brick buildings of Fort Thomas for more waiting. Henry called Lil and told her that he would not be home that night, since they were being held over for the blood tests until doctors were available. A noncom, officious and aloof, herded the men to a barracks where they were issued a towel, razor, blanket, and sheets. Henry was miserable. The bed was uncomfortable, the barracks were stark, and the lights and noises were strange and cold. Some of the men hollered, groaned, and talked in their sleep.

Their first army-life dawn found the new men being kept busy with classroom lectures. Night came again but there were no doctors until the following evening. Efficiently, the men were lined up and tested like pieces on an assembly line. The blood test made Henry feel dizzy, so he laid down on a bench for a while. The next two weeks would be his last days of civilian bliss for "the duration." The duration was an indefinite time, but at least it implied an end and that was what counted.

It was golden autumn. The fields and trees were lit with every shade of brown and yellow and sunlight. The sweet, hard little pears were falling in the yard from the tree that towered over the house and scented the air around it, driving the bees crazy. Henry spent the time around home with the family and went hunting with Fred Erhardt several times. The boys always compared their skills in that eternal "who was the best shot competition." They also were grateful for the extra food.

Meanwhile, Grandmother Elizabeth was just beginning to forgive Henry for a childhood misadventure. As a little "Junge Henni" he took up a slingshot and was aiming at a butterfly among the peony bushes. In the excitement of the hunt, he did not realize that the perfect shot missed the butterfly but did find another target---Grandma leaning over the peonies.

Lil noticed that Henry became quieter as the days slipped away until he had to report back to Fort Thomas for duty. His Dad went with him. Henry Sr. would work seven days a week at Carlton Machine during the war.

(Side Note) Grandma Elizabeth was very superstitious, like most Germans, and scolded Lil for having her picture taken out in the yard proudly holding the cat which belonged to Aunt Marie Strecker. It was bad luck. When Henry went to training camp, he received a letter from Lillian detailing how she had prepared a special dinner with just enough meat for their Dad, Henry Strecker Sr. It was veal at .98 a lb— a fortune in those days. She set it out on the table upstairs and went downstairs to get something. As

she walked back upstairs, the cat raced down past her. Her heart sank. What greeted her upstairs was the table with an empty plate cleaned off by the cat.

Lillian worked at Auto-Lite and rode the bus to work. During the last few months of the war, she heard people on the bus, discussing how they hoped the war would last longer so they could pay off more of their bills. That made her extremely angry. All she wanted was for her brothers along with so many neighbors to come home to Finneytown soon.

In a low-roofed brick barracks, Hank and the other men were given army clothes and shoes. The world outside the barracks, across the walls, lit by neon and street lights, seemed far away now. Several overnight passes were given and "the girls" as everyone called his sisters, Lillian and Ellen, took a picture of him the first day he came home in uniform. He posed, smiling, in his overseas cap, dark army tunic, light trousers, and dress shoes in the corner of their small upstairs living room for the picture. In camp, he was just another soldier, but at home, he was The Soldier.

Within a week, Henry was jouncing along in a convoy of buses which poured regularly out of the Fort to the train station and destination boot camp. The men trooped onto the train and it rolled away from the farm fields of Ohio, on through Washington, Baltimore, and up to New York. The East coast was crystallized under a silent and lonely snow.

In New Jersey, the train ground sleepily to a halt. Hank

felt the wet air and looked out at the landscape—a dull expanse of grey sand under a blanket of grey mist that blindly felt its way around the scrawny outlines of a few gnarled oaks and scrubby pines. Row on row of bleak orderly streets stretched across the sand peninsula. A cold sea swelled and rolled off the shore in the distance. This was Sandy Hook, Fort Hancock.

Here, as part of the 113th Infantry Regiment, Hank would receive thirteen weeks of basic training, learning the art of being a rifleman, how to use mortars, machine guns, and the M-1 rifle, along with finer arts like peeling potatoes and digging ditches. Henry also gained a New Jersey accent. (The 113th, a New Jersey national guard unit dated from 1775 and became known as "the orphan" regiment since it was used extensively to fill replacements in other outfits after D-Day.

The tents where the men were quartered had plywood sides and canvas tops with five cots and a small stove in the center. Reveille was at 6:00am and Hank would tumble out in the icy black air, dress hastily (sometimes simply pulling his rubbers on without his shoes) and line up in the company street for roll call. Someone would hold a flashlight for the officer to call the roll while the men grunted their presence in the darkness. They washed and shaved for breakfast at 7:00 and then went back to the tents to make up their beds. Close order drills, calisthenics, and an occasional trek out on the sand dunes learning how to use a compass, filled the morning.

There were more shots from the medical department—sometimes two at a time and the result was some very sore arms. After one battery of shots, the men could hardly get their heavy wool great coats off, and just had to shake them onto the floor.

Henry pulled coal duty only once. The tent stoves were supplied by a fuel box at the end of each company street and in turn this box had to be kept stocked. A truck trundled around the camp and a detail of men shoveled the coal into the boxes. There was always the danger that a tent stove would overheat and catch fire. Civil war soldiers in winter quarters had been plagued by the same hazard. Only one tent burned at Sandy Hook while Hank was there.

It was a difficult toughening up process and a severe head cold Henry had on leaving home worsened. He went to the aid station finally and wound-up spending Christmas in the hospital. "The girls," his sisters, who never forgot anyone away from home, sent him cookies and rolls of tobacco. That afternoon he felt well enough to eat the traditional dinner of turkey, cranberries, and mashed potatoes but later he broke into a fever again and felt much worse for having the heavy meal.

At about this time, a huge Italian guy on the ward who had gauze stuffed up his nose to stop his chronic nose bleeds, managed to untie his hands and pull the long strips of gauze out, cursing all the while. It created quite a bit of excitement on the ward. The next night a nurse came around carrying a tray with several two ounce glasses of a brown liquid. One was for Henry. All he could think of was the customary shots of castor oil given at home and he asked the nurse if this would taste like castor oil.

"Oh no, no, no," the nurse pleasantly assured him, and Henry gulped the evil looking brew. It tasted terrible. After almost a week in the ward, Henry still felt sick to his stomach, but was checked out. He relapsed, but by then it was good-bye to boot camp. From Sandy Hook, he would remember most of all the time-revered army tradition for lousy food. After basic, they were shipped to a series of camps along the eastern coast.

Camp Woodbine, thirty miles out of Atlantic City, was their first assignment. Here they would patrol the beaches and it wasn't long before the men walking the boardwalks were calling themselves, "The dirty sons of the beaches." Henry and another fellow were assigned the beach patrol from dusk until 1:00 a.m. There were no lights on the lonely stretch of coast and Henry liked to watch between the planks at the water below as it burst into a green phosphor-lit spray against the piers.

As a night time guard, Henry didn't have to rise until 1:30 for dinner. In the afternoon there was time to read, write, polish shoes, clean up, and then supper, followed by the night patrol. "The boards" of Atlantic City, jammed with people and throbbing with music at the Steel Pier were temptingly off limits to the men on patrol.

Here, the company was blessed with a very good cook

and consequently good food, including hand-made cinnamon rolls, bread, eggs, oatmeal, and even apple pie.

There was also CPO duty which was nice to pull. A fellow on this duty just laid around headquarters and slept, read, or smoked away the time. He was a replacement in case someone on guard duty got sick.

Mortar practice continued in the misty swamps and woods. One day the target was across a stretch of water. Hank was loading the shells and soon the crew was having too much fun just lobbing the shells without bothering to aim each shot. No one noticed the base plate sinking in the mud. The sinking finally changed the mortars range and as one shell soared up, it burst against a tree limb above the men and hurtled back down.

Luckily, only one man was injured by the flying metal. This was a sobering lesson and the aiming and firing was resumed with all due seriousness. Later, ironically, this same mortar squad won a trophy for best aim in a contest at the camp.

One night was marred by a tragic accident. An Italian, Sacco, was patrolling the beach with a sergeant in a jeep when they hit an iron railing after a sharp turn on the boardwalk. The jeep flipped over the railing and Sacco, the driver, was killed. The sergeant hobbled around on crutches for a long time.

After Thanksgiving, Hank slipped a letter into a mailbox on a desolate stretch of Beach near Seehouse City. No sooner had he done so, than he began to wonder how often the mail was actually picked up from such a lonely stretch of sand in November. The letter reached Cincinnati, with his Thanksgiving menu in it, six months later.

The home on North Bend Road was outside the five-hundred mile radius which limited leave-taking, but that did not stop Henry from taking three day passes home. Once he was riding the rails back to camp and when he reported in, was told that his leave had been extended. (His sister Lil had received a call at home — but Henry was already on the train by then). Undaunted, Hank left camp and immediately caught a train home for his additional leave. Three days later, the outfit was moved closer to Atlantic City for more patrol duty at Absecon, New Jersey.

Passes were sometimes given out on weekends, but several men just took off whether they had leave or not — until Captain Leeman found out. Several men were missing without passes one night so the sergeant was ordered to sound the alarm — by banging on a huge iron wagon wheel hung between two posts. Everyone tumbled out for the roll call and M.P.'s were dispatched to collect the missing men who were punished by being confined to their section.

The barracks could get pretty lively. A sergeant coming in from night patrol shot his rifle over the sleeping men in the barracks to hit a duffle bag hanging at the other end of the room. Some clothes and a mess kit were ruined by the shot but luckily none of the snoring sleepers were hit. Another night, a fellow in the neighboring barracks fired through the windows to knock out a light bulb above a man who was sitting up in bed reading. The light and window were shattered.

Next, someone broke one of the light-green glass shades that hung from the barracks ceiling. It was disposed of—tossed in the bushes behind the barracks. Finally, a sergeant called Buck came in drunk one night and punched his fist through several panels in the back door of the barracks.

Of course, that Saturday, the captain called a surprise inspection. He stomped in with the lieutenant behind him toward the hastily repaired back door. The first thing he noticed was the broken glass lamp shade in the weeds. He wanted an explanation for that.

The shattered light bulb inside caught his eye and heightened his anger. Who did it? He wanted to know. There were several seconds of eternal silence before someone piped up about the eager "sharpshooter" in the next barracks.

With blood pressure so high it fairly blew out of his ears, Captain Leeman roared at the lieutenant to "go get that gun." The rifle was ceremoniously brought back, whereupon the captain grabbed it and furiously slammed it over a bedpost so hard that the barrel was bent. He then left in such a fury that he forgot to yell about the patched-up back door. One of the fellows picked up the rifle, hoping to fix it, but it was beyond repair. The irony of it all was not lost on Henry. Captain Leeman never did cure his men.

Drunk again, Buck was in his bunk and decided to shoot

his rifle out a back window late at night. Sgt. Laplan heard the shot and was sure that deer hunters were trespassing in the woods. He called his men (Hank included) from their beds and drove them in his car for several hours through the woods. It was the dead of night. Everyone was groggy, tired, upset—and in the know about who "the hunter" was—but no one would tell the sergeant.

Then there was Charlie. He was an older guy who always had money for liquor but not for shoe polish or anything else he could "borrow" from another guy. Hank was assigned to Charlie's tent, so he heard a lot about the sorties at the Black Cat café. On a binge one night, Charlie was cracked over the head with a two by four. The hospital patched him up so he could go right back to drinking. As Charlie marched in front of Hank on night patrol, Henry would get a constant whiff of alcohol from his breath. The dark woods surrounded the road and only the sound of Charlie scratching the top of his soup pan helmet broke the silent night. The scab on his head itched but he couldn't reach it. It made Henry laugh.

The camp had a lone sentry box with three glass-windowed sides, a door side, a stove, and a telephone. As the camp was being emptied, Henry pulled guard duty there. At night with the lights out and nothing to do in the small hut, the man on sentry duty could take the phone off the hook while the men in the barracks would do the same and put a radio near the mouthpiece of the phone so the guard outside could at least listen to music to pass the time.

When it was finally time to leave the camp for good, there was a party (another binge.) Their marching cadence, set to an old drinking tune, was lived up to by some of the men. It went:

"Drunk last night, drunk the night before,

Gonna' get drunk tonight, like I never got drunk before,

For when I'm drunk, I'm as happy as can be, for I am a member of the One One Three.

Glorious, glorious, one keg of beer for the four of us, Glory be there are no more of us,

So the four of us will drink it all alone."

The cooks baked jelly-filled cupcakes and there was a lot of drinking going on. When the binge got going, someone broke the stove pipe with a bat while another battle group splintered a good wooden table into kindling. The non-coms waged a jelly cupcake war in their barracks. At the Black Cat Café, another group allegedly wrecked the bathroom sinks. To seek reparation, the angry owner found Captain Leeman, who stormed back at the man, "What do you want from these men—blood?" The owner blustered, "You've got the toughest bunch out of the 80,000 men here." Word of the confrontation filtered down through the grapevine and made "the toughest bunch" feel pretty good.

* * * * *

The next camp was Tuckahoe. Henry was issued an old rifle with a rust pitted barrel and would spend hours trying to clean it for weekly inspection. The inspecting officer would take one look down the old barrel and bark, "Clean that thing UP, soldier." Another fellow had a new, easy-to-clean rifle and was often rewarded at inspections with "Give that man a pass." One time the guy got a pass, then got drunk, and lost the pass before he could use it.

The 113th left its mark on Tuckahoe too and "tore the place up" on leaving. At the party, Charlie got drunk and chased Joe around the meat cutting blocks in the kitchen with a cleaver. Joe was yelling, "I never did nothin' to you, Charlie, whad' da' ya' wanna do that for…" and running for his life. When Charlie got drunk, he didn't know or care what he was doing and after it was over, he didn't remember.

* * * * *

The next stop was Fort Dix. Here's what happened there in Hank's own words: "Word went around about the first week in March that we were going to be taken off beach patrol. The coast guard was going to take over complete coastal patrol. On the morning of March 6, we got up, had roll call, washed, and shaved, and then ate breakfast. After breakfast, we rolled up our mattresses, folded our iron cots, and stacked them up at the front end of our barracks, cleaned out our foot lockers, and packed our clothes and other personal effects in our barracks bags.

"With no place to sit down, we walked to the woods

along the driveway leading from Camp Tuckahoe to the road. There we lounged around on the ground among the trees waiting for the trucks to pick us up. It seemed like we waited for hours. We had some C rations, which were almost better than nothing for dinner. The men were getting impatient and wanted to get going. Around two o'clock the trucks came, and we piled on. It was up to Fort Dix. We arrived at Ft. Dix in the evening in time for supper.

"After supper we were assigned to our tent area. We were to stay in perambur (pyramidal) tents, five men to a tent. The beds were the same type of iron cots with a mattress. While at Ft. Dix, we went on hikes most of the time to keep us occupied. There was an airfield there and, being interested in planes, I spent a few evenings looking at the planes. (Hank got chills when he climbed onto a fierce looking fighter plane in the dusk, peered in at all of the controls and saw that the seat was plain, un-cushioned, cold metal.)

"There were fellows from the air force learning to take off and land B-17 bombers on the field. A friend and I decided we would like to take a ride in one of the bombers. We inquired at the field office to see if we could go up. They informed us that their fellows were just learning, and it would be rather dangerous, so they wouldn't let us. We were quite disappointed. We were at Ft. Dix about a week when we received orders to board a train for Camp Pickett, Virginia. The train ride to Pickett was much fun."

Spring was coming and the days were cool, sunny, and

pleasant. The first green of spring—gold—was sprouting and the bushes were struggling to green out along the train tracks.

One night at Camp Pickett, someone rolled a coke bottle down the main aisle of the barracks. It shattered against an iron cot leg and lay there until someone stepped on it barefoot in the dark. Another ruckus.

At Boston's Camp Miles Standish, Hank was put on KP duty and helped eye four or five huge cans of potatoes until two in the morning—but there was a reward. The cooks gave each man a steak, a quart of milk, and some pie before they trudged off to bed.

* * * * *

Time came to move on to England and more camps, prior to the invasion. It took all day for five thousand men to board the English Cunard liner, RMS Scythia. Henry put his gear in the forward part of the ship at the water level stage. Everyone was given a typed paper listing rules on board ship. One rule informed the landlubbers that they would not be excused for disobeying a rule, even if they didn't know it was a rule. That evening, Hank watched the tugs push the liner out while the engines revved up. They headed up the coast to New York for a rendezvous with a convoy which included a small aircraft carrier.

The food aboard was terrible. Burnt wheat was served as coffee and the bread was not fully baked. The only thing that the English were able to serve that was any good was tea. Once, a large pan of sausages was spilled onto the deck, which was filthy. They were scooped up anyway and promptly served.

The first night at sea, Hank was supposed to sleep in a forecastle (foc's'le) hammock, but it was miserable, so he wandered to a wooden bench in what had been the dance hall, located toward the middle of the ship, where the rolling motion wasn't so noticeable.

The soldiers got dizzy, sick, and retched anywhere and everywhere. The decks and hatchways were slippery from vomit. The sea was high, but Hank didn't get sick until the third day out. After four days, the ship was finally cleaned up. It was bitter cold and everyone was bundled in overcoats during the thirteen-day crossing.

The Limey sailors doubled as anti-aircraft gunners. No enemy subs were sighted on the trip but at times Hank could lean over the iron rail of the deck and seemed to feel depth charges going off below. Another time, he spotted nurse sharks swimming alongside the ship. They sailed up the Irish Channel and disembarked at Liverpool. The Queen Elizabeth, her decks lined with thousands of soldiers, was also in the harbor.

Many Liverpool streets were bombed into ruins. By train the outfit was taken to Ilfracombe in Devonshire, to a tent city on the heights above Bristol Bay. Here they would toughen up. Overhead, Wellingtons and P-47's practiced dropping what looked like white flour sacks into the bay.

The homes of Devonshire were quaint and charming, the roads narrow and small. It was a beautiful countryside and the GI's enjoyed tramping down the narrow lanes.

Then came the invasion on D-Day. Replacements were needed immediately, and the vast tent city was evacuated. After riding in trucks for five hours, the men found that the situation was Army normal—all fouled up—orders had been mixed up and the 113th was not scheduled to leave for several more weeks.

Finally, the orders to move came and the tent city disappeared within hours, leaving only the barren grassy hills, the sky, and the sea. There was a half hour ride to a train station and a four hour wait for the train. Everyone was anxious to move, and Hank started to joke sardonically with the guys around him, "Where's that train. I sure wish it would come ... I'll go out of my mind if it doesn't come soon." This was emphasized by his imitation of the radio drama voices of the day which were so popular at the time.

A lieutenant listened to him with a smile on his face, as the rain dripped steadily from their helmets.

The train finally slid in on the black wet rails, the men boarded, and were off to Plymouth. Past three o'clock the next day, they arrived in the port city. Cans of peaches, (a real luxury item), gas masks, rifles, and suits with a stinking, wax-like, gas proof covering were issued. The suits were too tight-fitting and uncomfortable when pulled on over their uniforms.

The seas were still very high and rough as they made the

trip to Normandy in a converted Channel pleasure boat. They left the marshalling yard at Plymouth on June 13th. On June 19 at 0300, the attack on Cherbourg began. June 19th was also Henry's twenty-second birthday. He spent it on the channel boat waiting.

Silvery barrage balloons floated above the ships and in the water floated wreckage and the swollen bodies of dead men. Two weeks after D-Day, the beaches were still not cleared. After weeks of waiting, it was time for the replacements from the 113th Infantry to go ashore onto Utah Beach on June 20th. The rough sea made boarding the LCI difficult, but the driver assured the men that he would put them right onto the sand of the beach. That man let the ramp down in five feet of churning water. The riflemen sloshed through it, struggling to keep their duffle bags and equipment. Wallets left in pockets were thoroughly wet by the time they reached the beach with soggy money and papers.

Tufted grass grew along the line of fortifications and grey, sandy hillocks. Pup tents were set up on the beach and the men waited. A bomber attack was expected but did not materialize. They could hear the heavy guns up the peninsula blasting Cherbourg. On June 25th, Cherbourg fell after hours of fire from 155 mm. "long toms," anti-tank units, and air attacks.

Hank was moved right to the front, into riddled Cherbourg as a replacement for the Fourth Infantry Division, 12th Infantry Regiment, First Battalion, Company C on June 26, 1944. It seemed natural for Hank to pair up with

Vince Pilla. He had known Vince from the 113th in New Jersey.

Vince had been in another platoon. He was a Brooklyn Italian who had told Henry that, at home, if his father didn't like a meal his mother fixed, the old man would just dump the whole table over. Vince had two sisters—one married and a brother who was in the New York City police department. Wherever a dice game started, there was Vince.

He lost a little and won a little, but was always willing to give the shirt off his back to anyone who needed it.

Years later, Henry wrote this: "On June 26, 1944, I entered Cherbourg, France. The Fourth and Ninth Infantry Divisions had just taken the town. I was to be a replacement in Company C, 1st Battalion, 12th Regiment of the Fourth "Ivy" Division. The whole Fourth Infantry Division had lost a large number of men from June 6th to June 26th.

"On this day, I met Richard Engle from Columbus, Ohio. He was the BAR man, and I was appointed his assistant. I was twenty-two and he was eighteen. He proceeded to tell me what my job was with the Browning Automatic Rifle. I was to have a cartridge belt holding two twenty round magazines of .30 caliber ammunition. I would take over the BAR if something happened to him. As soon as we left Cherbourg, we were attacking the Germans again."

As Henry wrote: "Just behind the front line, they pitched their tent for two days. They were positioned in a serene, grassy field with a stream flowing down its center and several nice shade trees." Rations were passed around that evening and Henry ate five of the chocolate bars. The next morning, he woke up dizzy from the chocolate but with a handful of francs, he and Vince stopped at a nearby farmhouse and bought half a canteen of French cognac. It was as clear as water and very strong. Along the road, young pigs were eating the dead cattle that lay everywhere. German equipment also littered the roads.

Riding through Edmondeville in trucks, the men could smell burned cattle, or perhaps villagers as it was rumored. The first sergeant took out some photos of his family and talked about how he wanted to buy a farm in the country for his wife and children after the war. Near the front they jumped off the trucks, moved up a hill, and waited behind their first hedgerow. Rifles and the dead of the 83rd Infantry Division were strewn all over the fields and in foxholes. An 83rd man wandered up and moaned, "We're all shot up. There's nobody left." The man walked away.

Henry did wonder from the looks of things how anyone really could be left alive. Two German machine gunners lay dead in their hole where a mortar shell had killed them. Another man lay dead with his side blown open. The Germans began to send in intermittent fire. As Hank wrote, "We were suffering heavy casualties of dead and wounded. It was a nerve shattering experience. We crossed a concrete footbridge under shell fire and there was a dead man laying there." Henry dashed over the small bridge, praying that no shells would come in as he ran. Midway he had to jump over the dead soldier sprawled on the crossing. He felt alert and noticed every detail and at the same time felt like he saw nothing around him.

Fear was blinding. The bursting shells made him feel sure that he didn't have a chance—and this was only the beginning. Hank wrote, "That evening I said to some of the men around me, 'Did you see that dead German lying back there by the footbridge!' They straightened me out really quick. 'That was no German, that was one of our men.' I was so nervous, I guess I expected to see only dead Germans. Everyone was so covered with mud; it was hard to tell who was who."

They dug in behind a hedgerow and in the afternoon, moved to another field full of dead 83rd men, into a gully over a creek and dug in behind another hedgerow. The awful stench of the dead cattle in the next field, legs stiff and bloated, wafted over them. White tape used to mark mined areas, was strewn around. It began to drizzle.

About nine or ten the next morning, they moved out, spaced in a single file formation, across the next field to a garden. The air was hot and clear, a nice sunny day. They crawled carefully through the garden. The sweat was pouring off Hank in hot streams on his face. No shots were being fired, but he was scared. Through a gate, they crossed into the next field which sloped up to a tree-lined hedgerow and slogged through the hay while their two scouts started over the hedge and down the hill. Their first sergeant was rattling on to his new men all the way, trying to calm them with small talk... "come on boys... let's go ... keep moving..." He was talking for their benefit. He understood.

Hank peered over the next hedgerow and saw the scouts prowling carefully down near two small knocked out German tanks. Nestled among the trees in the valley, far off down the hill, was a barn with its Dutch style doors open. Suddenly a sharp burst of machine gun fire sprayed the slope. The scouts raced for cover and the GI's began to fire toward the farmyard. The enemy was nowhere in sight. Hank took his bayonet off and jabbed it into the hedgerow. He thought the Germans might see it in the sunlight. At any rate, the Germans knew the Americans were behind that hedgerow. Hank fired several shots at the barn door and window. He urged Vince to stick his rifle over the hedgerow and fire a few shots too, but Vince was sitting tight and low and wouldn't budge.

A machine gun was hurriedly set up at the corner of the hedgerow and blasted the farm position while German shells, machine gun bursts, and bullets flew toward them. Suddenly the lieutenant yelled, "Break 'er down." The machine gun was dismantled and the guys raced back for cover while a hail of bullets and fiery tracers lit the air above them. The sergeant was still yelling encouragement and it charged Henry with confidence as he headed for cover.

He was hot, exhausted, and thirsty when they dived into their old position at the gully. He reached for his canteen, but it was gone—lost. The tight, impregnated oily suit made breathing hard, his heart pounded in his ears. To relieve the pressure, he wanted to throw away the new tobacco pouch that bulged in his pocket. It was impossible, so he just emptied the tobacco out and put back the pouch with his pipe.

It was getting dark and outpost guards were needed. When the sergeant called out "Strecker"—Hank felt like dropping dead. A BAR (Browning Automatic Rifle) man, another fellow, and Hank went halfway up the slope and dug in. At dusk, German shells howled into the main compound by the creek. Hank watched the shells spray sparks and sheets of flames and with each explosion he was grateful to be on that outpost duty. All night the three men strained their eyes in the darkness, watching the front.

At dawn they moved down to the farmhouse. Hank wasn't really sure what had happened, but the lieutenant and the first sergeant were both gone. He was told that during the night the lieutenant had shot the first sergeant—they had never gotten along too well—but the sergeant was always easy to get along with and a nice guy.

The farm yard was completely plowed up by shells. The dead 83rd Division men lay everywhere, their faces chalk grey and yellow and caked with mud. Near the well in the yard, Henry found a German pistol which he very carefully picked up. The Germans were well known for booby trapping tactics, but the Sauer 7.65 was just lying there.

Broken tree branches from the shelling were lying everywhere and in one of the farmhouse rooms, the Germans had left two sides of beef hanging. They smelled as rotten as the dead cattle in the fields. Mine sweepers were scanning the farm lane when some tanks rumbled up. One rolled onto a mine and its tread blew off. The disgruntled crew got out to wait for a repair outfit. A jeep pulled onto the lane and hit a Teller (plate) mine with about twelve pounds of dynamite in it. At the explosion, Hank wheeled around to see tires and metal flying almost seventy-five feet into the air in a billow of smoke.

The men dug in behind the nearest hedgerow and waited. Hank sat holding his breath as the German artillery went off in the distance and came hurtling in on them, louder and louder. The bursting shells made a ripping, crackling sound. After each explosion, Hank would thank God and think, "Well, that was another one that didn't get me." A farm building started to burn. The shelling stopped.

Water was needed so Hank and Vince were sent back to the water supply truck. They were both very thirsty and drank their fill before heading up the road with their precious cargo. On their trek, P-47's with 500 lb. bombs on their wings, circled ominously. Hank began to frantically look around for a sheltered area just in case, but no place looked safe. It was impossible to tell where those shells might land, and he didn't want one in his lap.

When the war birds moved on to attack a road junction up ahead, he was relieved.

Back at the farm, only one cup of water was doled out to each thirsty man.

That night, one of the men moved ahead too far and found himself hiding inside a corn shock, listening as the unsuspecting Germans chatted around him. Luckily, they pulled out and he escaped. When he came back safe the next day, everyone was happy to see him. He had a family and had been given up for lost. In the fighting several weeks later, he was killed.

The outfit moved onto the next hedgerow, past an apple orchard. The trees were greening out beautifully. Beneath the sprawling, budding trees was a sizeable herd of dead cows.

The Fourth Infantry Division had fought up the Cherbourg Peninsula with heavy losses and turned south toward Orglandes, while the Germans had almost a month to strengthen their hedgerow lines. Since the area was swampy and rough, the 12th Infantry Regiment had to attack along a German-picked corridor. With deadly artillery and mortar fire, the 17th SS Panzer Division and the 6th Paratroop Regiment, made the Americans pay heavily for every hedgerow. On one day, the Division only advanced four hundred yards. It took a week to gain four miles and bad weather prevented air support. The infantry had to go it alone.

The German 88's were pouring shells on them one night when Hank was in a foxhole with a fellow named Treauex, who started to cry and pray out loud. Hank was praying too—but not out loud. Suddenly seized by fear, Treauex jumped up to run but Hank was able to pull him down and talk him out of it. Shells were bursting everywhere so their best bet was to stay hunkered down low.

During an intense fight in Normandy, Hank and Vince were lying several feet apart on a slope. Suddenly Vince hollered that he was hit by a machine gun burst that had ripped into his back. Hank stayed put and Vince began to cuss him out, yelling that if he ever got hit, Vince would be sure to just leave him. Finally, Hank gave in and crawled over to Vince. It turned out that all the cursing had been over nothing more than a barbed wire scratch on Vince's back.

The 12th Infantry Regiment broke the German line and took LaMoisentrie, Neuville, Sainteny, Les Forges, Roffeville, La Roserie, and LaMaugerie, advancing to Periers.

Hackley, a guy from Harpers Ferry, Virginia, was walking a little ahead of Hank on the other side of the road along with a BAR man, a skinny Youngstown boy, wearing a very large, heavy cartridge belt. With a steady clackety-clank, two vehicles which sounded like half-tracks, were approaching them. It was dark. Near Hank and Hackley, the vehicles stopped. The men held their breath. A figure in the first truck stood up silhouetted against the sky and asked where the front was — in German. Without a word, Hackley shot the man. The column of GI's scrambled wildly up the hedge walls with their hands torn and scraped by the thorny wild roses. The bazooka man blasted the vehicle as Hank and the BAR man made it over the hedge and through a field.

During the night, they dug in and watched a thatched house roof burn. Hank was so exhausted he didn't care if he did get killed. A white-haired, blonde goateed sergeant told him, "Get over there and watch. The Germans are right on the other side of the hedgerow!" Hank crawled over and sat watching through the hole blown in the hedge by a shell. The sergeant trooped back and told Hank to get out of that hole—or he might become a target. He moved.

The next day the sergeant went down to the German truck. It had been pulled to the side of the road and the men checked to see if it was mined. There was a small red cross painted on the truck. GI's were walking at the other end of the lane. Apparently, the other German truck was able to back out of the lane the night before when the shooting started.

Once, they hiked all night, stopping only ten minutes every hour for a break. The march was exhausting, and the men would lean wearily against trees or just plop down along the road for the rest intervals. As morning broke, they stopped at an apple orchard for a half hour. Hank's feet never hurt so bad before or after, but his legs were loosening up with every step. The column plodded on. They advanced up another farm lane which was lined neatly with beautiful sentinel-like poplars. The GI's turned left and dug in along the hedgerows and in the fields. That afternoon they took off again and by evening had moved about three or four hedgerows ahead before they dug in for the night.

Guards were posted since the Germans were near. Hank took his turn at guard, posted near a wooden gate behind a hedge. Hank could hear Germans in a farmhouse just ahead of them. They were having a noisy party with music, singing, and a lot of hollering and laughing. They were probably also on patrols in the area. Suddenly Hank froze. Listening in the darkness he could hear someone crawling up the other side of the hedge.

His rifle was too unwieldy to aim around the gate, so he slowly and carefully pulled out his pistol, clicked off the safety, aimed directly at the advancing figure's head, ordered halt, and asked for the password. From the darkness came a whispered reply. The password. A GI passed through the gate and Hank breathed a sigh of relief. He stayed at his post until his time was up, woke another man to take his place, and went to his hole for a few welcome hours of sleep.

The next day fresh baked bread came from supply. It had been loaded into burlap bags and trucked to the front. Eagerly, Hank sank his teeth into the beautiful crust of the soft white bread. What luck! The bread crust was full of miserable grit and sand from the bags. It irritated his teeth to no end as it settled between his gums—but the men ate it nonetheless. They were hungry and the bread was good.

Mail—a huge morale builder—was also brought up

that morning. After eating, they moved forward down a hill and up another to a hedge and a little barn with a decorated tile roof. Two scouts, the sergeant, BAR man, and Hank started to crawl over the hedge. Machine gun fire stopped them.

As Henry wrote later, "The Germans opened up on us with their machine guns. Their bullets would kick up the tile on the roof of the shed behind us as they aimed their machine guns back and forth. We were going over the hedgerow and I was behind Richard Engle. As he climbed up on the hedgerow, the Germans opened up with their machine guns. Engle flipped back off the hedgerow like you see in the movies, right on his back. A couple of us immediately tried to find out where he was hit. It didn't take long. The bullet entered the top of his shoulder and came out his back near his shoulder blade. The bullet came out sideways, leaving an opening in his back in the shape of a .31 caliber German bullet.

"I checked his backpack after cutting the straps loose from his shoulders. The bullet entered the pack and hit a button on his raincoat. It bent the point on the bullet in an L shape. I gave it to him and said, "Here's a souvenir, keep it." He grinned and clutched it in his hand.

"I could see where his lung was moving up and down through the hole in his back. There was blood coming from his mouth. The aide men arrived shortly and carried him away. I said good-bye to him not knowing whether he would live or die.

"I took the BAR and began firing back at the Germans. Not used to firing the weapon, I held my finger on the trigger until it stopped firing. The BAR spit those bullets out so fast, I thought it had jammed on me. I finally realized the magazine was empty. I then removed the empty and jammed a full one in the receiver. I finally had the gun so hot I couldn't touch the barrel. I thought, I better oil this thing before it does jam. We carried a small can of oil in a pouch on our cartridge belt. I poured some on the receiver and it was like pouring water on a hot stove. They brought up more ammo from the rear and I had to start reloading my magazines from the bandoleers of new cartridges.

"Our bazooka man fired a round into the corner of the farmhouse on the other side of the hedgerow near the Germans. After a while, the firing subsided. "Then the fighting continued. German machine gun fire ripped and splintered the tile off the quaint barn roof all afternoon as the Americans returned fire. At one point, Hank saw a cloud of thick black smoke after an explosion near the hedge. He was afraid it would be a German mortar adding to their misery, but later realized it must have been a hand thrown "potato masher" going off. Hank took his BAR, laid it over the hedge and kept firing bursts until it jammed. The magazine was empty before he realized it. He re-loaded and kept firing into the gate across the field, the farmhouse, and the trees.

A bazooka fired at the house and took a corner out

of it—but the machine gun kept chattering. One of the younger GI's, Irving J. Russell, was scampering back and forth along the line passing out ammunition, which was running low. He raised up too high when the Germans sprayed along the top of the hedge. A bullet cut through his helmet and liner, grazing his forehead as he joked it off saying, "The Germans can't kill an ole' Irishman."

That afternoon an air strike was called into their sector. The GI's had to roll out an air identification marker. It was a pink, luminescent paper about ten feet long and thirty-six inches wide, rolled in a canvas bag. They would have to spread it out and pin it down with rocks to indicate to the pilots where their own lines were located. There was only one hitch. No one could find it. There was a frantic search—and it was laid out as the air attack started—too close to the American lines for comfort.

Finally, some weapons platoon men crept across the field toward the farmhouse, peered into a hole by the house, lobbed a grenade into it, and hurriedly ducked. Two German machine gunners were in that hole. One was wounded and the other lost several toes in the explosion. The GI's dragged them out and hustled them across the field. One particularly jaded American held his .45 against the wounded German's head and threatened him angrily. Those two Germans had held the company up all day.

When they got back behind the hedge, the scene changed. The Americans became solicitous and politely offered the Germans cigarettes and a light before the aide man took them to the rear and out of the war. The outfit stayed there for the night and dug in.

It was quiet. At dusk, Lieutenant English took a patrol up around the farmhouse. He never came back. A sniper had shot him between the eyes. Everyone missed him. The grey-haired, grey-eyed Lieutenant was very well liked.

K-rations were passed around the next morning for breakfast and the men started over the hedge again. No shots were fired. They walked down a hill to another farmhouse. At dusk, the men were slogging along on each side of a lane hemmed in by tall poplars and hedges. Tanks had been through and shot up one of the farmhouses. There was a wounded French girl inside and the aide men took her back for medical help.

The outfit continued down through the fields, came to a road and were slogging up it when some of the men heard motorcycles coming. They took cover in the woods and before long two motorcycles roared into sight from around the hillside. Machine gun fire knocked the two Germans from their cycles. One was already dead as the anxious GI's pulled them off the road. The other was severely shot up and stunned. His tunic was soaking wet with blood and there didn't seem to be much hope for him. An ominous clanking grew steadily and threatening louder. It was a German tank.

As it clattered around the corner, a bazooka team fired on it and the creaking black panzer exploded into flames. This sent the platoon scurrying back through the woods.

They thought the sheet of flames was the tank firing on them and they weren't about to face it with rifles.

The outfit was pulled west of Carentan to assemble for the needed breakthrough in the Southern sector. They were tented on a hill until then. One morning while shaving and washing up, a dogfight started in the summer sky above them. It would be the only one Hank saw overseas. The machine guns of the planes purred high above the startled infantrymen. Through the clouds, Hank saw a P-47 chasing a Messerschmitt and losing ground fast. A second Messerschmitt flipped into a tailspin and crashed hard into an open field about a mile and a half down the slope. (Later "Stars and Stripes" reported that the downed Luftwaffe pilot had been a noted German major.)

The men on the ground froze, when what appeared to be another Messerschmitt started to dive straight at them. A .40 caliber AA gun was quickly manned but a sergeant yelled "Hold fire." Too late. The men on a nearby machine gun didn't hear him and began pouring rounds into the swooping fighter. The plane crash landed on their flank. Too late the machine gunners realized their prey was a Spitfire.

* * * * *

On July 25, 1944, the brakes were off again. Three thousand heavy bombers opened the attack. The 12th Infantry Regiment with armor in the rear moved in behind the 8th

Regiment and mopped up pockets of enemy resistance. Henry wrote, "Our next objective was St. Lo. It was in the hedgerow country of Normandy. We were suffering heavy casualties of killed and wounded. The next day we attacked again going through the same maniacal, mind bending, mind shattering shooting and killing. We took St. Lo on July 26, 1944."

Under German fire, the regiment moved eight miles to the north of Le Bourg, on the night of July 28th.

Ernie Pyle was with the 4th Infantry Division for the breakthrough, as he wrote: "I went with the infantry because it is my old love, and because I suspected the tanks, being spectacular, might smother the credit due the infantry. I teamed up with the 4th Infantry Division since it was in the middle of the forward three and spearheading the attack."

In a recorded telephone conversation, Feldmarschall von Kluge, Commander in Chief West, German Army said, "Villedieu, springboard for the east and south as well as Avranches, is the anchor point for Brittany, has to be held under all circumstances or else has to be recaptured."

On August 1st, the spearheading 12th Infantry Regiment captured Villedieu. Allied air attacks were heavily strafing the German retreat. Hank passed the wreck of a German staff car. There were many on the road. Moving south, they took the heights over St. Pois after heavy fighting and on August 6, were relieved for bivouac and a much needed rest at Brecey.

As they marched back, newsmen filmed them trudging along the road. Some men were passing out from fatigue, some lay down in the road exhausted, and everyone was dirty, unshaved, and sweating. The heavy BAR was being passed along the line. Every ten minutes another man took his turn carrying it. Vince was begging Hank to take his turn for him. Hank said no and as usual, Vince began cursing him out. Among other things, Vince joked that he was thirty-six and hoped that when Hank was thirty-six, he would be on crutches. They were buddies—always joking—and good friends too.

* * * * *

On August 7th at 0400 the heaviest counter attack in Western France hit the Allies. The area around Mortain was smashed hardest and the 12th Infantry Regiment was rushed in to help the 30th Division which could not keep enemy armor from entering St. Barthelemy and Mortain. The 120th Infantry Regiment was cut off on Hill 314 east of Mortain. On the 8th of August, the 12th's Third Battalion joined the 117th Infantry on the right flank of the fray at St. Barthelemy. A two thousand meter juncture separated the 117th on the north and the 119th and 120th on the south. It was the job of the 12th Regiment to fill the gap and relieve a trapped battalion of the 120th by driving south into Mortain. The First and Second Battalion were to advance ten miles along the road. Four miles of this was marsh country exposed to German artillery and mortar fire. The 2nd Battalion was to take the crossroads. Hank marched with the First Battalion headed for Mortain. Before they got two thousand yards, German mortar, and small arms fire from a hill-studded area west of Mortain stopped them.

Company B fought all morning on the first day of the attack. Company C moved along in front of a hedgerow, past a farmhouse and orchard. A dead German in a camouflage uniform lay under the trees. Hank thought he was probably a sniper.

Then they moved into a clean mown field, walking in double file formation, and at its corner turned. Two tanks were lined up nearby. There were German machine gunners and soldiers behind the opposite hedgerow in a dip, who had the GI's in plain sight, but had not opened fire as the Americans plodded past. The high hedgerow may have prevented them from seeing up the hill. At any rate, they could have picked off the unsuspecting Americans, but did not. As the GI's reached the corner of the hedgerow, the Germans finally opened fire. Hank and his outfit raced down to the opposite hedge.

There was a lane and another hedge flanking it. Both of these hedges had a gaping shell hole blown through them and the men found themselves exposed to a murderous German cross fire. One man fell wounded. Firing was so heavy and their location in the open field was so bad that a withdrawal to the morning position was ordered. Hank and Vince crossed the hedge at its corner and went back down two fields to their former position on the road near the farmhouse. An excited artillery captain tromped up, curious to know how the fracas was going and told them to "get the hell" back up front where they were needed. Hank and Vince went back up.

A hail of bullets was coming in from different directions and withdrawal was again ordered. Hank was grazed in the leg by a bullet that ricocheted off another fellow's rifle. One of the tanks had pulled down into the next field and was knocked out. Hank saw a tanker, clutching his wound, running back. This time Hank and Vince crossed one field and got in behind the hedge. An officer trudged up. They didn't recognize him. It was Colonel Jackson. His objective was to take a road up ahead that would cut another pocket of Germans off. He wanted to see what was holding things up and asked the two what was going on. Hank answered, "Nothing right now. We couldn't get anything and they told us to withdraw. We've been up there twice and each time we had to withdraw, but if you want me to, I'll go up there again."

Abruptly the Colonel asked Henry if he "wanted to lead those men up there." Just then a wounded man on a litter was being hurried past them and Hank added, "That's the way we're coming back." The Colonel didn't say another word. He could see what the situation was for himself and left.

Armor piercing shells whirling end over end began

pounding into the ground. Hank and Vince couldn't figure out just what kind of shells they were for a while. Armor piercing shells didn't burst. The Germans were firing them over their own lines from a Tiger tank dug in and camouflaged high up on the opposite hill. Luckily, they overshot and fell beyond the company line. The rattle of machine guns and the heavy shelling went on all day.

In the next field, Hank watched a line of B Company men huddled together on the open ground behind the hedge. It wasn't a safe thing to do, and they soon found out the hard way. Artillery support was called for and the long toms began firing. Their range was off, and they fell in short—onto the B Company line. A single shell knocked out seven of the GI's who weren't dug in. Hank helped carry one of them back on a litter. The boy was spattered with dirt and mud and lay on his stomach with his left leg twisted completely around and laid very neatly, wide open. The wound was burnt black—and not bleeding. The man was groaning, and Hank didn't think he would live.

That evening they moved into a farmyard across a hedge near an orchard. Near the house in the garden, Hank and Vince made a comfortable foxhole with a stout wooden door for a covering, then ate a ration supper and spiced their cheese with some small red onions which they had scrounged up from the garden. Someone came along and ordered them to move down and dig in along the same hedge that the Germans were on—except that they were on the other side of it and down further. They started digging and then were told to pull back. The ground was rocky and hard by the hedge. Someone "borrowed" a pick from the farm, so Hank and Vince were wearily chipping out a new hole when the outfit was again ordered back to the farm.

At this point they were angry, mad, and exhausted, but tromped back. Incredibly, they were once more ordered back to the hedge. It was too much. Bundles of kindling were piled high in the yard, so Hank and Vince pulled some of them down to build a barricade-like protection around their hole for the night. They were too exhausted to dig another hole and desperately needed to sleep. No one seemed to care whether the Germans came or not. No one posted guards, and if the Germans had come, they could have easily overrun the snoring Americans.

Hank woke from his dozing at the crack of a shot in the night and then heard some yelling in French. An old woman had been walking down the farm road and was mistaken for a German. A GI had taken a shot at her and missed, so she was busy cursing him out. Later, German bombers soared over, dropped flares, and circled back to hit the American artillery emplacements. When they left, the silence of the night finally settled over the front at Mortain.

That morning, Hank and Vince were roused from their sleep and asked why they hadn't dug in along the hedge-row. When Hank replied that they were just too tired, the questioning ended. K-rations were handed around and the men were told to eat quickly, the attack would resume. It made Henry feel sick. Another day of hard fighting was expected. They waited all morning before the order to move out. No resistance was encountered as they walked down behind the abandoned German position and saw the trampled grass where the German infantry had been. The Tiger tank sat deserted.

Effective fire from it had kept the company off the road a whole day. Across the hedge and a lane on the opposite hill was a small hedgerow enclosed field with ten or eleven smoldering, knocked out American tanks. They had rolled into the field through a back gate and once inside had been neatly picked off by the German Tiger.

On a hill in front of this field was another farm house. The day before some of the B Company men had advanced toward it when they saw GI's in the yard and were startled at being fired upon. The GI's at the farmhouse may have been prisoners with German guards firing at the company or Germans in American uniforms.

Behind the Tiger tank was a small road and across this was a Sherman tank with a German cross painted on it. It was common for the Germans to use a captured vehicle when they could since their supplies of fuel were low. It was then that Hank saw the first and last dead civilian that he could recall—an old French woman lying face down in a ditch along the road. He always wondered if it was the same woman who had come down the road the night before. The heavy firing early in the night may have killed her.

They moved past the Tiger into another open field. Hank picked up a German helmet since he had been looking for one the right size to send home. He examined it and saw a piece of flesh stuck in the top of it. It would not be shipped home. The tarred black road—their objective—was reached and crossed. They dug in beyond it, but the Germans did not counter-attack. The desperately fought-over road was no longer needed.

The Germans could be heard pulling out in trucks. They had bought the time they needed to get out. Company C stayed in position that day and K-rations were given out. On the next day they moved on down the small lane.

* * * * *

The fighting from August 9th through the 12th had been some of the hardest so far.

The SS Adolph Hitler Panzer Division, along with twenty battalions of artillery, other panzer units, good observation points, sunken roads for maneuvering, and great incentive for the Vaterland, had stymied the 12th Infantry Regiment. It had been the Germans last chance to stop the Allies in France and they lost. On August 12th the 4th Infantry Division took all objectives, destroyed six German tanks, and relieved the 120th Infantry. A German counter-attack failed, and the 12th Regiment pushed on.

Trucks were transporting the company when shelling started again. Hank was looking out of the back of the truck when they began to pick up speed and a jeep behind them took a direct hit. It didn't take long for the men on the truck to realize they were in a hot spot. They yelled for the driver to stop, jumped out, and dug in along the road.

During one dusk in Normandy, a German plane — maybe a medium bomber, flew over. The GI's started to dive under the cover of a nearby truck until the driver scrambled out of the cab, yelling that the truck was full of dynamite. Everyone dashed to get away from the truck as an AA ground crew began firing. Hank watched the bright tracers arc up in the night sky and fall short of the path of the plane.

On August 17th they were moved to Carrouges to fill a hole in the line of the Falaise Gap. Reinforcements joined them. Along the way, Hank's buddy, Vince bought some good strong cognac. Smiles, pretty mademoiselles, and an endless supply of wine welcomed the Americans.

* * * * *

The march to Paris was swift. Over the evening of August 23, 1944, the Regiment drove a hundred and sixty five miles by truck to capture bridges by Corbeil. It was raining and the roads were slippery. German planes swept over the convoy. By dawn the trucks were rolling over the plains toward Paris. The 12th Infantry Regiment and the 2nd French Armored Division moved ahead, and Hank could see the Eiffel Tower in the distance late that afternoon. Two miles out, the 12th was slowed up to allow the French to enter the city first. Between Versailles and Paris, the French met resistance.

The company spent that night in a field. It was cold and the men didn't have blankets with them, so Hank put his raincoat over himself and curled up to keep warm.

On August 25, 1944, the 12th Infantry Regiment moved north on the Boulevard D'Orleans and under the Port D'Italie, reaching the Notre Dame Cathedral at noon. The trucks stopped and Hank jumped out with many of the other men to visit the cathedral. The inside was very dark, Gothic, and inspiring.

Hank rode in the third truck in the first American convoy to enter the liberated city. Thousands of cheering Parisians lined the streets and throngs packed the sidewalks and street curbs. People were handing bottles of wine to the GI's. The men had three or four bottles each under their seats in the truck. That afternoon, Vince warned Hank to go easy on the wine — or it would really hit him hard later. A French girl struck up a "conversation" with Hank and when he mentioned Charles de Gaulle she burst out with "Non, non, non … Il est Chorl D' Gohl." Next, he tried to get across to her that his name was Henry Strecker, but she cut him off again with "Non, non non, AnRI Streck-AIR."

She was bubbling her French almost as fast as a German machine gun could fire and Hank gave up. An exuberant Frenchman invited Hank to a bar and gave him a martini in a large water glass. Outside, again the Frenchman pressed another glassful on him.

The outfit was moving out by this time and the lieutenant had to send someone to get Henry. The trucks rolled on and by seven o'clock the alcohol hit him. He almost fell off the truck, but his buddies grabbed him by his belt and held onto him. Twice they rolled over the Seine bridges and by 11:00 pm were quartered in a building. Hank went to sleep on a hard, plain bench. The night was interrupted by a Frenchman trying to appropriate a rifle. Another Frenchman had caught him in the act and that led to a lot of high-pitched French hollering. The helmet Hank left on the truck was stolen during the night.

The next morning, the wine hangover really set in for Hank. No breakfast, water, or coffee was issued. In the afternoon, quarters were moved to Vincennes Park and the motor pool finally appeared with blanket rolls. A hot dinner was fixed by the company kitchen and the pup tents were pitched. For the first time in weeks, Hank had a chance to pull his boots off and wash his sore feet—in his helmet of course. By the time he finished washing and shaving the dead skin off his feet, the ground was white around him.

There remained a gruesome reminder in the Park that the war was not over. The covered bodies of FFI men shot by the Germans as they left the city, were lined up in rows awaiting burial.

There was time to loaf around, and a guard roster was

made up. Hank was out on duty for a stretch of time. Some men went out to see the Cathedral. That evening the Germans dropped incendiary bombs on the city and Hank was sitting near his tent when the Luftwaffe droned overhead. White balls of fire from the explosions bounded toward him and his heart pounded with fright. He knew he'd never be able to escape the ball-like flames. They blazed out and fell short. Then the Germans dive bombed. A lone Frenchman shot at the dark silhouetted planes with his rifle.

The following day they were marking time at the Park, when a French girl came in and started "chatting" with Henry. He spoke no French and she spoke no English, but she took a fancy to him. The other fellows egged her on and fixed up a date for Hank despite his protests. With lots of gesturing at her wristwatch, the matchmakers showed the girl that she should come back at 6:00. He was trapped. On guard duty that evening, he completely forgot about his date and when he did remember, had to rush to clean up and gulp down supper. The girl had not forgotten. There she was at the gate—on time.

It was a sunny, pleasant evening and they took a walk on a neatly manicured, grassy mall. When they sat down on a park bench, a GI sitting nearby started to give Hank menacing glances. Hank wondered if the interloper was acquainted with the girl too. Henry was getting mentally prepared to take care of the intruder if he tried anything, but the irate fellow strolled away. They sat on the bench until about 11:00 pm when sirens began to wail.

Bombers began thundering over the city and Hank didn't know where to go—but the quick-thinking mademoiselle grabbed his hand and led him to the shelter of a towering church while the bombs thudded and boomed in a distant part of the city. When all was quiet, Hank walked the girl home. The streets were lonely and fog-shrouded. It drizzled.

Alone, Hank started back to Vincennes Park. He couldn't help but think that the citizens, oppressed for so many years, were acting out understandably angry. Many rode through the streets clinging to the sides of their cars with guns at the ready; others whizzed along on bicycles, guns in hand.

On these same black streets, they had shot lone German sentries. Crowds had gathered to fight in the streets, wreak vengeance on Germans and collaborators, or just watch and yell. As he felt his way down the dark Paris blue streets, he was afraid that someone trigger-happy might mistake him for a die-hard German sniper. Just in case, he pulled out his pistol and held it ready. Near the park, a convoy of trucks rumbled past him. The men in the trucks were grumbling and Henry thought that it was just another unlucky outfit having to pull out. He was wrong. Vincennes Park was deserted. The company, his tent, his BAR, and his equipment were all gone. Only the kitchen remained.

All that next day, he rode on top of a truck and not until nightfall, when the kitchen was being set up could he ask anyone when the next jeep was going up to his company. He wanted to be on it. The following morning a jeep rolled away with breakfast for the company, but without Hank. No one had bothered to wake him. Not until the following evening, did he get a lift to the front along with the supper delivery. Henry explained his absence to his sergeant who didn't seem to think anything of it or say anything about it to him. Not until the war was over, did Hank know that he had been marked as three days AWOL and because of this was almost refused a good conduct ribbon.

Over two hundred GI's had deserted and disappeared in Paris. Most—like Henry—moved on. He just wanted to get home—fighting the war and ending it was the only way he could do that.

Ernie Pyle summed up Paris perhaps the best when he wrote: "As usual, those Americans most deserving of seeing Paris, will be the last ones to see it if they ever do. By that I mean the fighting soldiers. Only one infantry regiment and one reconnaissance outfit of Americans actually came into Paris, and they passed on the city quickly and went on with their war.

"The first ones in the city to stay were such non-fighters as the psychological warfare and civil affairs people, public-relations men, and correspondents. I heard more than one rear-echelon soldier say he felt a little ashamed to be getting all the grateful cheers and kisses for the liberation of Paris when the guys who broke the German army and opened the way for Paris to be free were still out there fighting without benefits of kisses or applause. But that's the way things are in this world."

* * * * *

The rest of northern France was covered by foot and truck. The Germans put up no large-scale fight during this part of the retreat but left pockets of resistance troops and counter-attacked to slow down the Allied drive before being mopped up.

The Company trudged past a sign which marked the road to Soissons. It reminded Hank of some of the World War I stories he had read. They were nearing the area that Remarque had described so realistically in *"All Quiet on the Western Front"*. The next town taken was St. Quentin which had been a scene of action in the last three European wars. Now the outfit would turn east to the Meuse River at Fumay on the Belgium border. They were in the foothills of the Ardennes—a plateau of thick forest and deep terrain. Here the Germans would have ideal defensive positions.

On September 6th the Company crossed into Belgium. The people of the little villages stood in their doorways or along the roads to greet the Americans, hand out hard boiled eggs, and glasses of cognac and wine. They waved and smiled and were very friendly. One woman handed out freshly made waffles from in front of her house.

The Americans were right on the German heels when they reached the river Meuse. The Germans had blown the bridge out and the GI's had to climb down a steep ladder with their unwieldy equipment to reach the water near a bridge pier that still rose from the river. They crossed the water over a narrow, rickety, plank bridge. It was a pretty hair-raising walk and Hank expected the wobbly planks to collapse and throw them into the river at any minute. They dug in for the night after a safe crossing.

For the next seven days they would forge ahead and cut the German retreat off at LaRoche. Hostile territory was being rapidly approached. Their next route through Schlierback and St. Vith was within the 1940 German border. The people here spoke German for the most part and some sympathized with the Germans. Anxiously, the GI's would ask where the Germans were. The answer was always a hand gesture indicating—right up ahead of you. Hank expected some action soon. He knew the Germans would not retreat forever.

Two or three days after crossing into Belgium, Vince and Hank were heating their C-rations when Vince picked up his hot tin and burned his thumb. Vince was griping and beefing about his finger, so Hank just had to tease him with, "I hope you don't get any sleep all night." There was a lot of good-natured kidding around between the two of them. They liked to act like they hated each other—just like brothers.

Outside of a small village, a woman handed Vince and Hank each a raw egg. As they walked up the road, Vince put his egg in his front shirt pocket. Hank heard a shell explode up ahead and knew that the Germans would let them have it soon. They walked down the hill to a valley in single file lines on each side of the road. The Germans had an eighty-eight down on a hillside, out of sight. Vince wearily shifted his rifle to his other shoulder. Crunch. The rifle strap crushed the egg in his pocket. He pulled it out, a gooey, slippery mess and threw it away, cussing all the while about his rotten luck. Taking a lesson, Hank decided to eat his egg before it suffered a similar fate. Besides, he was hungry and needed a lift. He cracked the top off and for the first time in his life ate a raw egg.

They passed houses along the road and entered another town. On the edge of the village, they could see where a shell had hit a tree and torn out part of the road pavement. The men were ordered to spread out, form squads, and head off along the road. The German artillery spotters were good. Hank called a warning to Vince, "Get the heck away from me. There's no sense in both of us getting hit by the same shot." He heard the ominous thud of three artillery pieces being fired in succession and added, "Sounds like we're going to get some artillery, Vince."

Vince was climbing over a fence just a little ahead of Hank when the first shell slammed in. Hank saw Vince knocked down and roll back—hit by the shrapnel. There was no place to take cover. Hank jumped into a three-inch deep cow path near the fence and hugged the ground. He started to raise himself up when another shell slammed in. The third shell rocked the ground as he tried to get up again.

The shell holes had torn into the field in a triangular pattern. On the fourth try, Hank got up and scrambled down the hill. The firing stopped. Hank felt as if he had been hit—punched hard in the side. He could hear Vince hollering his name. The German shelling resumed. Hank ran up the road and down along a lane to a line of trees. Hank checked his jacket. There was an L-shaped tear in it where shrapnel had entered, hit his magazine case, nicking it, and spun back out. Some American 4.2 chemical mortars were set up and a barrage was laid in on the Germans who were down in the woods. They were withdrawing.

Hank went back when the shelling stopped, and he realized what had happened. It had happened so fast. Some of the other men were hit and Hank saw them, along with Vince, being taken away on a jeep. Hank desperately wished that he could be with Vince.

Morale was at rock bottom. Hank turned away and went into a garage with Hackley. Both of them felt sick and wondered if they were going to get killed. Hackley kept threatening to go back to the aid station himself—he just couldn't take it anymore.

Hank felt overwhelmed himself and couldn't say anything to Hackley, except that he wouldn't stop him if he left. They went toward the next town and slept in a hayloft that night, taking turns on guard at the door. No one got much sleep that night. They were too upset.

The Germans were moving, and Hank listened to their trucks pull out the rest of the night. There was also the sound of glass windows in the town shattering and tinkling onto the streets. Some people were still in their homes.

The next morning, they left, tramping up the hill alongside the road. In the afternoon they came to a woods where they could peer down to a town and river below them. They would have to stop. The Germans were in the town and on the opposite hill. Attack formation was ordered. German mortars dropped in on them, killing one man and wounding four or five others, holding everything up until they were ordered to rush the town. The Germans withdrew. On the hill, the GI's found a half track and a camouflaged .88 with shells left behind by the Germans. It had been fired at them earlier on the road. The men stayed in a pine woods that night and took some grain-shocks from a nearby field to cover up. One of the men on guard duty that night accidentally planted his boot across another man's neck and the sleeping, startled man let out a blood curdling scream. He said he thought the Germans were trying to choke him to death.

At daybreak, Hank asked about Vince. He was told that Vince had lost his leg and probably gone into shock and died the night he was wounded. It was hard to comprehend that his buddy was gone, and it really hurt. They had been through so much together for so long. (Years later Hank's face would stream with tears from the trauma at the thought of not being able help Vince when he was hit and never seeing him again.)

They took off again. The German border was crossed. The company hiked until dark and finally dug in. The next day they would hit the Siegfried Line—the German defensive West Wall built with a maximum of propaganda and a minimum of concrete.

The 4th Infantry Division men felt their way toward the Line, a wall of cement dragon teeth and strategically placed pillboxes. It stretched a mile to the east amid rough terrain, thick woods, deep stream beds, steep slopes, and rounded hills. (The 4th Infantry Division has the distinction of being the first Allied unit to cross into Germany and cross the Siegfried Line. That happened on September 11, 1944.)

The outfit moved up a hill toward a high, pine covered ridge. Pine trees also flanked them to the right. The day was gloomy and overcast and the Lieutenant asked the BAR men to take the lead. Germans must be up ahead somewhere in the gray mist. The men walked slowly in formation, about ten feet apart. Two scouts led, followed by the BAR men, assistant, officer, riflemen, and platoon guide who brought up the rear and saw that everyone was in line. There was no hurry. The Line was just up ahead. No one knew what to expect or how bad it would be, but Hank was sure that it couldn't be much worse than what they had already been through.

The soft earth was damp and not a twig snapped under their feet. It was eerily quiet. The field was knee high with red-gold grass. It reminded Hank of the buffalo grass in the fields at home — except that here, barbed wire and regularly spaced tank obstacles scarred it. The tank obstacles loomed in a rigid line. They were pyramids of three metal shafts stacked and bolted together.

The company moved down a knoll toward a pillbox. Shadows appeared—Germans were moving through the pines. Moments of silence passed and then bullets started to spatter in on them. The GI's couldn't see where the Germans were and started to creep stealthily up to their first pillbox on the Siegfried Line. It was empty. Oddly enough, the next pillbox they found was also empty. The men continued to follow the line of pillboxes which were staggered along the hill—placed strategically so that they could cover each other.

A lone German was shot down at the next pillbox they encountered. Dusk came.

The men were deployed around the concrete bunker and dug in among the saplings and high weeds. Their shovels hit rock. Spades chinked and clanked on the stony ground as the men grumbled and whispered. German machine gunners fired a burst over their heads.

That night, Hank slept in a shallow hole with a rock sharply pillowing itself in the middle of his back. The next day they withdrew a little. Volunteer scouts were called for, but Hank didn't step forward. He had been upfront long enough to know better. Penetration of the Siegfried Line was accomplished that day by the Second Battalion.

Late in the afternoon, they moved forward again and dug in for the night. The ground was rocky here too and the Germans sent shells in all night. Hank threw up a dirt barricade on one side of his hole.

In the morning, they pushed forward. In a small, long

wooden house on stilts, they found sixty Germans with an officer, ready to surrender. Some of the Americans found good pocket knives that the Germans had left in the hut, while Hank and several others were detailed to lead the German prisoners to the rear. An interrogating officer lined the Germans up and asked them to count off in lines of "elf"—eleven to a row. By dinner, Hank was back with the company again.

The lieutenant wanted to take the next pillbox, so they crawled out and approached it from the rear where there was a single thick metal door. The Germans sat tight inside while the GI's had a free hand to try to dislodge them. A pole charge failed to work, and they were unable to blow the door off. Shells started lobbing in to the right and left. These were American mortars, smoking and tearing up the sod dangerously close. Luckily they were duds. American observers must have sent in smoke shells to spot the place and then fired duds. The men had no success at the pillbox and moved back to their holes. That night was quiet. Dawn found them holding their position and nothing was happening.

That evening a group went up to check the pillbox. It was vacated. The weary GI's settled down for a quiet night in the concrete bunker. Then came a knocking at the door. It was opened to a courteous German soldier who queried, "Wasser?" (water). The startled Joes countered with "What!" and that was enough. The equally astonished German dropped his Jerry can of water and scurried off into the darkness. The bunker door was slammed shut and the GI's stayed.

At daybreak, a German tank could be heard roaring closer. An attack was expected.

Support was radioed. Rifles would not stop a panzer. Sure enough, the German tank pulled up to the door of the pillbox, leveled its .80 mm cannon and blasted three armor piercing shells into it. Two TD's with .75's clattered up the hill. Their lieutenant jumped down and crawled ahead, pointing out the target. His drivers, riding high, with clear vision, aimed and loaded. Their first shell rocked the German panzer and its crew started to clamber out of the hatch when another shot slammed into them. They tumbled back down, and a third shell jolted the panzer.

The GI's in the pillbox ran out alive but scared and shook up. They looked like ghosts, powdered with white cement dust. Blood trickled from cuts where pieces of flying concrete had blown into their faces. The shells had roared around the inside walls like marbles in a can. Miraculously, no one was hurt badly. One of the men, Sebastian, came out with his eyes rolling wildly, he yelled hysterically, "The dirty s.o.b's sent those armor piercing shells right in on us."

He was a nervous wreck and Hank never saw him after that. The Germans in the panzer were killed by the concussion of the TD shells. The tank destroyers left the area.

That afternoon, the Germans counter-attacked. It was one of the few times that they did this in such strength. It was September 19th and the SS Regiment Deutschland, supported by Mark IV tanks and heavy artillery had moved up for the attack. Hank listened to the pong of .80 mm. mortars being loaded, followed by their soaring up and plummeting down with a loud swish. The shells were falling behind him. To the right were trees, and on the left was a knoll that dipped down behind them. Hank kept watching that left flank and his front. It wouldn't be hard for the enemy to creep forward right there. German infantrymen were already moving up among the trees trying to get in close for some good shots.

Joe Juartez stood behind a tree and calmly picked them off like clay pigeons as they bobbed about from tree to tree, letting themselves be exposed or helping a fallen comrade. The firing became heavier all this time and suddenly Hank noticed that the other fellow in his hole, Roy Rock, had simply disappeared. No one, for that matter, was in sight and Hank became more dazed as he realized he was watching the flank and front alone.

Finally, he decided to pull back to another hole before he was cut off and captured. Mortars were exploding around his new position. "Where is everybody!" kept spinning through his head. Then he remembered his shaving kit. It was in that hole at the front he had just left. And the more he thought about it, the more he wanted it. The heck with all that firing, no lousy Kraut was going to get his good shaving kit, so he went back up for it.

To his amazement, he found not only his kit, but his buddy Roy, back in the hole. A handful of fellows were nearby. Hank never could figure out where they had gone to or come back from. Actually, several units on the thinly spread line had withdrawn to re-organize before the advance.

Hank stood in his foxhole with a sergeant and Roy, and they kept firing. One German was popping his head up closer and closer to one of the GI foxholes in front of Hank. The wide-eyed replacements in the hole were too scared to look out and spot the German so the other men were yelling to them about the menacing Kraut who was inching closer. They were urged to throw a grenade directly out in front of them. The faint reply came from the new men that they didn't have any. Next, they were told to just stick their weapons over the top of their hole and fire point blank. No go. The men were frozen and just wanted to surrender, but the fellows behind them kept urging them to hold fast.

They had been fighting since Normandy and weren't about to surrender. Someone lobbed an undetonated grenade up to them. It fell short of their hole, and they would not reach out to get it. Meanwhile, the Germans inched still closer behind a small dirt road that rose like a small plateau in front of their foxhole. Finally, Hank took aim with his BAR just as the German bobbed up again and fired three rounds, forcing the heavy rifle to kick up against his shoulder hard. The first shot ripped the dirt on the edge of the replacements hole and the next clipped the foliage. The German dropped down at the same time. Hank wasn't sure if he had hit the man—but he never raised up again.

The BAR jammed and Hank thought it was broken. He took it apart to clean it thoroughly, along with the magazines. Pine needles and dirt had clogged it. It would have been a drastic time if the Germans had charged them just then.

Mortar fire was called in and the confusion and racket began again. The sun was shining but it was blotted out for an instant when an .80 mm. shell hit a tree and knocked the top out of it right behind the men. Artillery was quickly told to knock off their fire. It was too short. Support was radioed and two medium tanks with .37 mm. cannon and three machine guns came rattling up on the left flank to spray the area with gunfire while the GI's moved forward. A tanker popped out of his hatch and asked if everything was OK now. The infantrymen answered yes, and the tanks pulled back.

The Americans had really pounded the area. A German draped in a camouflage poncho slicker came running up to Hank and Roy with his hands up, eyes staring wildly. He was scared stiff and kept babbling "Kamerad … Kamerad!" They waved him on back into their lines. Hank found the German he had fired at dead in the underbrush. He had been hit so badly by the tanks that Henry couldn't tell if he had shot him or not. He took a large pocket watch and a cheap version of a .38 pistol made similar to a Lueger which he gave to Roy. Roy Rock was new with the outfit. This was his first real fight, and he was shook up.

They passed two Americans dead in their foxholes, and

as they advanced Roy walked up on a young German in a hole who was slumped over a machine gun. Warily, Roy, who was still jumpy, kept his rifle aimed at the German. Slowly the German opened his eyes. He was probably as terrified as Roy. Before a word could be said, Roy shot the man between the eyes. Fire fights were like that—there was no time to think or talk. The German could have shot Roy just as easily perhaps, but it was over and Roy was left stunned, almost crying.

They withdrew to their holes for the night, having broken the German line. Not until after the war, at Bamberg, would Hank, with others who had stood firm on that hill, be given the Bronze Star they earned.

The company was exhausted from holding the line and the next morning the 2nd Infantry Division, wearing the Indian head patch, moved up to relieve the 4th Infantry Division. The sound of TD's moving probably caused the Germans to fear an attack, so they began shelling. The GI's didn't even have time to dig foxholes. They tried to huddle around the TD's. One of the men jumped in on top of Hank and Roy, startling them, but what else could he do?

They settled down when a shell hit a large tree nearby. The shrapnel ricocheted down off the tree and hit their "guest" who was sprawled across Hank and Roy. He yelled that he was hit and headed back off the line. Hank and Roy had been spared only by his unexpected leap of fate.

At about noon, orders came for the 4th to pull off the line. The weary men shagged it back eagerly, while the new outfit jumped into the abandoned holes. Company C was only too happy to be put in reserve and dug in about a mile back. Rumors went around that the Germans were sending out nine or ten man night patrols and Hank expected a chance meeting with them, but luckily that never materialized.

Pup tents were set up, hot chow was served, and a canvas enclosed shower with overhead water pipes was set up in a field near several houses. The shower had wooden plank boards thrown on the ground as flooring, but they were pretty useless since the field was so muddy. In fourteen months oversees, Hank had one sponge bath in Normandy (out of his helmet, in an open field), this shower on the Siegfried Line, a shower before his furlough to Paris later, a bath in a real tub at a Paris hotel, and a shower in a tent while it was still cold in the spring. The combat infantry smelled earthy, like the ground they slept in, and their shirt collars became so heavily caked with grease and so brown from dirt and sweat, that they were as stiff as leather.

The Germans had a distinctive odor also. It was easy to tell if they had been around or were around — even if there was no sign of them — simply by the smell of their fuel oil which permeated their equipment and hung in the air even after they had left.

For about five nights in a row, Henry dreamed that the Germans were trying to catch him. In the first dream, he was walking through an open field and bending down to pick up some .32 bullets which were scattered there for his pistol when trucks pulled up. He didn't pay much attention to them until he noticed that they were Germans. When he started to run, they were chasing hot on his heels. A twenty foot gate sprang up in front of him, and as he scaled it he could almost feel the German bullets tearing into him. He kept going. He had climbed about five high gates like that, with the Germans still chasing him, when he woke up.

Another time he was dozing off in his foxhole and slipped into the last of his nightmares. Some fine, hairy shrub roots hung out like a web from the wall of the hole above him. He dreamed that a German with a net and bayonet was trying to catch and stab him. Hank was frantically trying to push the nightmare away when he woke up with the roots dangling above his face.

A new guy, Biandi, was made Sergeant. Combat infantry badges were given out. Long, blue and silver, showing a rifle and oak wreath, they were given to front line Infantry men only. It made Hank and the other veterans feel good. As Napoleon would say, "Give me a bolt of purple cloth and I will conquer the world."

For the rest of September, they stayed in a defensive position in the Schnee-Eifel forest building up to strength. On October 4, the Division was relieved, and the 12th Regiment went to Holzheim, Belgium.

Hank had been sleeping with a fellow named Pace. When Pace breathed, something in his nose made a loud clicking sound which he could not help. It nearly drove

Hank crazy at night. Pace also had high blood pressure and a bad heart. The man could hardly breathe if he stood next to a running motor. It never ceased to amaze Henry that the military would accept a man in such pitiful shape for front line duty.

* * * * *

The next battle came at Losheimergraben. The men called it Buffalo Bill Chalet after a hunting lodge near the town. The Company was ordered to get ready to go into the attack and dug in for the night close to their jump-off point. They were in the Ardennes and many of the men felt demoralized. Treaeux, who had served in Iceland, was sent to the front without a furlough home and by then, he was suffering mentally. It only takes one man in this condition to demoralize other men. That night, Treaeux made a cross out of sticks and planted it on Hank's foxhole. Hank thought it was a pretty sick joke, got mad, pulled it up and threw it away.

At dawn, they started through the woods under radio silence to approach the town. Houses were in sight when enemy fire pinned them down. One of the scouts was hit and Second Lieutenant Alphonso Barracks crawled out, got the man's arms around his neck, and carried him back to the lines on his back. Hank admired the Lieutenant—like a good officer he never expected anyone to do what he wouldn't do himself.

They withdrew and re-formed. Artillery was called in, but it fell short. Luckily, no one was hit. Hackley was surprised and elated to see Hank alive. He had seen a shell burst over where Hank had been standing minutes earlier and was sure that Hank had been vaporized in the burst. The artillery stopped and the TD's were called in. The engineers started to clear a path for the TD's to roll up and fire into the houses ahead of them.

The Germans opened up and the clipped trees sifted sawdust down on them like a fine snow. Hank advanced to a hole near the houses and saw several explosions go off. He thought they must be booby traps until he saw something fly up in the air, hit a tree and fall again, exploding. It was a potato masher. In a foxhole, Hank found a German clutching a little white card — trying to surrender. The German moved his hand out to the edge of his hole toward his neatly arranged row of four uncapped potato mashers. Hank kept his aim on the man just in case he would try to grab one of the grenades he had been throwing. He waved the German to the rear and jumped into his hole for cover. Another German, this one dead, lay next to him as he grabbed the air cooled, magazine-fed machine gun mounted on a tri-pod. He fired the German machine gun into and between the houses until it overheated and jammed. Then he picked up his own weapon again.

During the attack, Irving Russell, the Irish guy, who in Normandy had joked that the Krauts couldn't kill an Irishman, tried to cross to the right of the road and was cut down by German machine gunners who were nested in the ditch along the lane.

Several German Luftwaffe men who had put up a good fight were taken prisoner. Hank carefully checked into one of the houses. As he entered the door, he was startled to see a figure standing inside. His first impulse was to fire his leveled rifle at the uniformed man. There was no time to say a word. As his eyes focused, Hank realized quickly that the person he was looking at was his own reflection in a full-length mirror on the wall.

They moved ahead and dug in beyond the town. The dragon teeth of the Siegfried Line could be seen winding down through the woods like a path. The following day they went back to the Chalet. It was a quaint, once fashionable hunting lodge. Upstairs, the bedrooms had beautiful shiny satin quilts. Some were bright blue, pink, orange, or red. Some were also missing. The Germans had taken the thick, warm quilts and used them in their foxholes. Across the hall from the bedrooms was a veranda which overlooked a field and view of the woods. This was the Jagdhaus, built around 1905 at Rocherath-Krinkelt. By the end of the war it was destroyed by artillery from both sides.

Patrols were sent out on reconnaissance around the line and Hank went out with four other men. They crept silently through the still terrain. It was uncomfortably silent and lonely. The next day, they needed a new sergeant and Knight, a young kid who had been AWOL a long time, was given his stripes back. Hank and the older men were once again disappointed.

Outposts were picked — one in a small valley between a point where two creeks flowed together and another higher up on the right bank of one of the streams. Hank was assigned to the position at the fork in the streams for a week. It was foggy, cold, and snowy, but hot chow was taken out to them. Then he was rotated to the base position for a week where he had time to write home. His next duty was at the stream bank.

One afternoon they were told there would be pillbox assault practice for an outfit nearby, along with the time it would begin, so they wouldn't think they were under attack. Hank decided to get some BAR practice while the exercise went on. When the mock siege began, the GI's at the outpost also began target practice. Hank aimed his shots at a bush across the stream. Not satisfied with this sport, Hank took some toilet tissue wrapped to a stick and began waving it wildly, pretending to surrender. That made the GI's at the stream outpost start to laugh and joke.

Meanwhile, the tanks on the other hill were advancing with flame throwers for the exercise. If anyone heard the outpost "skirmish" at least nothing was said about it. Hank's third and last outpost duty here was at the fork in the stream again. One grey morning, the two fellows who were with him decided to have some fun at his expense. He could hear them plotting near the tent and prepared for whatever was in store. When they started to chase him, he took off down the hill. They were going to try to pull his pants down and he knew it. In the ravine, Hank stopped short, whipped around, pulled out his pistol and fired one shot over the taller GI's shoulder. Game over. The pursuers froze in their tracks, then sulked away cussing at him for "doing a dirty thing like that to a man."

Late one evening, the sun was streaking through a beautiful cloud speckled sky when the men heard an odd bumping sound. Hank looked up and saw a winged black buzz bomb—a V-1 skimming across the blazing sunset with its fuselage glowing fierily.

Hank had never seen anything like it before. It was fascinating and eerie to think that this rocket was flying on its own to a target. In foggy weather, the V-1's could be heard going over and when the buzzing sound stopped, they knew they could expect to hear an explosion. It was scary. They could never tell if one might explode right over them once the engine noise stopped.

Word filtered through the ranks that Field Marshall Rommel had died of wounds received July 17th in Normandy when his staff car was hit. The general opinion was, "Well that's just one more we don't have to worry about anymore." Of course, later history revealed that the Field Marshall had been forced to commit suicide after being involved in the July 20th plot to kill Hitler.

* * * * *

On November 6, 1944, the 12th Infantry Regiment was ordered forty-five miles away to Zweifall, Germany. They would be attached as reserve to the 28th Infantry Division. The men were ordered to cover their holes carefully in this new landscape, since it was flat, low country and liable to strafing air strikes. They boarded trucks in the evening and rode all night in the rain, wind, and cold. During their trip, the orders were changed and delivered to the column commander, "The 12th Infantry will relieve the 109th Infantry (28th Infantry Division) in the front lines." The fate of the 12th was sealed.

It was a cold, wet dawn in the forest called Huertgenwald. Hank and his buddies jumped off their truck and sank knee-deep in mud and slop. No breakfast was given. They headed up the road and to the left for the attack. As they hurried past six 105 howitzers, the guns fired, almost knocking them off their feet. They sloshed off the road and into the woods to form for the advance. They would have to relieve the 28th, unit for unit, and rushed to swallow their rations. There was snow on the ground, and they hadn't gone very far when the firing started.

Colonel Luckett had no chance to organize his units or patrol the area. All afternoon they were pinned down in the deep snow. Shells hitting to the right and left tore up the trees. Sergeant Biandi got a flesh wound and had to be sent back. Everyone was happy for him. He would be out of it for a while. American artillery hit the ridge on the left, throwing up black funnels of smoke and shattering the trees. The company withdrew to their starting position for the night and were given hot chow.

In the morning, K-rations were passed around and they headed forward in a different direction—to the right. When they passed through some pines to open ground with several large trees, the Germans again pinned them down with machine gun fire. One of the GI's yelled as he was grazed across the chest while standing behind a tree. He was taken off the line. Hank started to dig in and got his hole about three inches deep when the German machine gunners fired over them. Hank dropped flat as bullets clipped the foliage inches above his head.

An artillery observer officer came up to help. The ground troops needed support in the dense, netherworld of the forest. The first round called in fell to the right. It was behind their lines. The officer upped the yardage, and another round came in. Hank thought it would hit directly in their lines but some of the other men guessed it was OK, so Hank didn't say anything. The lieutenant radioed, "Artillery fire for effect" and warned the riflemen to get down in their holes. There was the danger of shells bursting when they hit in the trees above the men and raining shrapnel on them.

Hanks' premonition was true. Twelve rounds fell into the company line. Several men were killed, and others injured, including the artillery officer and one of his men.

Later that afternoon, the German machine gun nest was spotted, and its two occupants surrendered to one of the men.

They dug in for the night. It was still raining—a miserable rain in the gloomy forest. There were no rations or blankets. The men covered with their raincoats, but that didn't keep rain out of their foxholes. In the morning, they waited and then withdrew to their original starting point, passing two men from their Company who had been killed. One was Staff Sergeant Harry R. Vandercar. On November 11, 1944, Harry helped an aide man pull "Pete" R. A. Holcomb to a road to be evacuated when German artillery hit hard. Harry quickly put himself over Holcomb to protect the injured man, when the third shell came in killing Staff Sergeant Vandercar. Sadly, he left behind a wife, Ida, and two daughters, one was only two years old and the other eleven months old.

After the war, a missing in action list was passed around to the men when they were in Bamberg. If anyone knew whether the men on the list were deceased, they were supposed to check their names off the list. Hank had to check off Harry's name. Incidentally they had both been replacements from the 113th Infantry. Staff Sergeant Vandercar was never forgotten, and his grown daughters, Lois and Elaine had him re-interred at Arlington National Cemetery. More information may be found at www.awon. org.—American WWII Orphans Network.

So many men Hank met were killed, wounded, transferred, or just seemed to vanish and he never saw them again. The 4th Infantry Division had 4,025 total casualties in the Hurtgen Forest (killed, wounded, and non-battle), our second highest casualty month of the war. 5,400 total casualties in June 1944 was the 4ID's worst casualty month.

The company then moved on trucks parallel to the German lines and dug in near some fields in a pine woods. Hot meals were promised but never arrived, and their foxholes were half dug when they were ordered to pack up and move on in the trucks. Up front, the Second Battalion had been surrounded for two days by the Germans and they had to be bailed out. Screaming Meemies were coming in on the front and right as the GI's hurried up a firebreak into a neighboring field where they dropped their coats in a heap. An artillery barrage began. Hank dove for cover and a piece of shrapnel about one fourth inch thick and steaming hotly ricocheted into his hole. He scrambled out and down the firebreak. No one was in sight, so he kept going and caught up with the company.

Dead Germans and Americans were stacked up all around the area, stiff and swollen. The men attacked to the right through an opening and into a pine thicket where they bumped into some tough Germans holding a firebreak. A hail of bullets clipped the tree branches down. Captain Whittkopf, commanding, was hit in the leg with a German egg grenade and refused to leave. He stayed on his feet for a long while but finally had to be taken back. Hank couldn't help but think it was because he stayed at the front so badly injured, that he lost his leg on the operating table. Other men were getting hit and had to go back past the rows of dead men to the bunkers. There were two lieutenants left in command—one of them was Barracks. They went around and told the men to rest and eat well for the next day when they were going to attack with the help of Company A to relieve men who were trapped by the enemy.

That night was miserable. Hank shared a hole with Lavey, who had a severe head cold and runny nose. They talked about getting home and out of the "crummy war." At dawn, eight or nine men of A Company crept quietly forward. They had no officers. Fresh snow covered the ground. Lieutenant Barracks led them silently through the pines to the right. Someone had a .30 caliber machine gun from weapons platoon and after a long search, the Germans were finally spotted. Someone yelled, "Let 'em have it" as the Germans made a run to escape. Two or three were captured. One was an old man in his sixties or seventies. Hank and the other men were so thirsty at this point that they were drinking the melted snow water they found in shell holes and ruts.

The entrapped Second Battalion was found and rescued. The hard hammered men ran toward Hank and the other men, tears streaming with relief at being released from their encirclement by the Germans. They had no medical supplies left and were sure they would be taken prisoner soon. The Germans started to send in armor piercing shells and tank shells exploded overhead, popping the tree tops off as if they were toothpicks. Four or five tree tops were knocked off in succession by one shell before it stopped.

The GI's stayed put that night and Company C men crawled into holes which had been dug earlier. Hank stumbled around in the dark looking for a place to sleep.

After the war Henry wrote: "The third time I was hit was about the time Ward Means got wounded. C Company with one platoon from A Company were attacking the Germans in the Huertgen Forest. Snow was falling on the morning of November 11, 1944. F and G Companies, Second Battalion, 12th Infantry Regiment were surrounded by Germans for about three days without food or medical supplies. It was do or die.

"Heavy weapons D Company set up two water-cooled .30 caliber machine guns and opened fire as we charged the Germans. We were yelling, "Let 'em have it" as we ran firing our weapons. The Germans were completely caught off guard. I took one German prisoner out of a foxhole. The rest ran off into the woods. I started telling the boys from F Company, lying in their foxholes, that we had broken through the German lines. They were extremely happy and some cried. I was out of water, so I started scooping up hands full of water out of holes torn in the ground by shell fragments. I felt like I owned that forest.

"The Germans started firing .88 mm armor piercing shells at us from a Tiger tank. The shells would hit the three to four-inch thick pine trees, snapping them off like toothpicks, three or four at a time. The tree trunks would snap in a cloud of steam-like vapor. I walked around like nothing could kill me. That feeling is very rare. Most of the time, you're groveling for a hole and praying you don't get hit. Also shaking like a leaf.

"It started getting dark and orders came up that we had to spend the night there. I started looking for a foxhole to crawl into. There was a dead German in a hole, and I grabbed him by the boots and started to pull him out, but the cloying smell stopped me. I did not think I could cope with that odor all night."

There was again, no water, no blankets, no food. Hank scooped water from a shell rut. "Ray Litterst occupied the next hole I came to. Ray invited me in to spend the night there. In the middle of the night, I woke up. Ice water was dripping down my neck. My feet were ice cold. I raised my head to look down at my feet. During the night it started to snow again, and my boots were covered with snow. The elements were taking their turn at giving us the works. At dawn we awoke with nothing to eat. Ray had an extra K-ration bitter sweet chocolate bar which he offered me. I appreciated this tremendously.

"After about three hours we received orders to withdraw from these positions. In the meantime, the Germans dropped a small mortar round on us, and it hit in the pine trees overhead, sending a shower of snow down on us. The shrapnel hit no one. I thought to myself, 'this does not look good.' They are probably zeroing in on us. As we moved on out, I saw a new Browning Automatic Rifle used as a support to hold up the front of a foxhole. I was tempted to change it for my old one, but second thoughts stopped me. Mine was tried and true."

Ammunition, blankets, and rifles were strewn on the ground. The Second Battalion men moved out first. "As we moved on out, we had to pass through some cleared ground. As we suspected, we had given them four hours to set up their artillery. The German artillery gunners were second to none. All hell broke loose. By this time in my fighting career, I became an old hand at finding the nearest hole or whatever to crawl into. I found a hole with four other fellows in it. We were in there like sardines.

"November 13th, suddenly I heard someone calling, 'Help me fellas.' I immediately recognized the voice of Ward "Brownie" Means. I hesitated momentarily, thinking I'm relatively safe in this hole. But then when you think if that was me out there, I would want someone to help me. I scrambled out of the hole and ran over to where Brownie was lying in the snow. There was an aide lieutenant with Ward, but that lieutenant wasn't much bigger than me. I only weighed about 135 lbs. but the good Lord turned our bodies into steel. I put Ward's left arm over my shoulder and the lieutenant took his right. Ward was hit in the back and couldn't walk. His left hand had the forefinger tip and thumb blown off. Blood was running down the front of my raincoat. If anyone saw you dragging someone at home here with a back injury like that, they would swear you were crazy. But the time was NOW.

"Smoke was curling up out of the shell holes around us. We had to drag Ward over two hundred yards of shell ploughed ground with rounds falling all around us. Only by the grace of God did we make it. We put Ward on a litter and the aide men carried him back. They shelled us for over an hour before they quit. They used zonal fire on us, not changing the settings on their guns.

"I crawled into another hole with two aide men in it. The one aide man asked if I was hit. Ward's blood was on my coat and there were two holes in it. I said no. I opened my raincoat and fatigue shirt. I was wearing a German pistol I had found in a farmyard back in Normandy. I had made a shoulder holster for it. A piece of shrapnel went through my coat and shirt, hit the pistol, tearing the holster, and gouging the steel butt, then bounced out, leaving two holes, two inches apart. I still have the pocket out of the shirt, which means nothing to anyone but my family and me. God had his hand over me again."

The metal had torn his shirt, hit the leather of his pistol case, and came out two inches below at his pocket, but had not left a scratch on him. A small square of shrapnel, apparently from one of the tree bursts, had hit him. Several days later, Hank noticed another shrapnel tear neatly cut into his gas mask. After Brownie left, Hank got into a hole with a New York boy. Henry was hungry and late in the afternoon went back to get his canteen, cartridge belt, and shovel from his old hole. The canteen had a shrapnel hole in it and the precious water was drained out. He grabbed the cartridge belt and found another canteen lying on the ground—empty. That evening he filled it up down at the Weisser Weh stream. The water was cold, clear, and fresh.

In the forest, only the Germans seemed to know where they were. That was how it usually was. The GI's weren't told where they were and where they were going.

During World War I, Father Duffy, chaplain of the Rainbow Division had said, "We don't know where we're going, but we're on our way." If the men chanced on a road sign or shop name, they might know where they were, but these opportunities were few and far between. Day followed day and week upon week blended into months. Hank noted later that his service time passed quickly and seemed a time apart from the rest of his life.

Hank and the man from New York lay in their wet hole, pulling sopping wet GI and German blankets over themselves. Steam rose from their bodies and the blankets. Their feet were cold, and their meager water supply was gone. That night Hank went to the stream again. The next afternoon he asked the other Joe to fetch the water but he was too afraid, so Hank went again. The trip to the stream for a canteen of water was risky so he scavenged around and found a five-gallon German water can with a hole torn in its top. He filled that and plodded back the fourth of a mile from the creek with it. Even this supply didn't last long—only a day. The cold made them thirstier. To avoid the German shelling, Henry would go before breakfast for the water.

That afternoon they moved to a hole less deep and closer to the Germans. A fellow from weapons platoon came up and got in the hole with them to talk. Shells started raining in and one hit the small tree on the edge of their hole—only three feet above them. The concussion was terrible. It forced Henry's mouth open and all the fluids in his sinuses were pounded into his mouth, leaving an odd taste. The man on top of him, who was coming from the aid station and looking for his platoon, was hit in the knee and began hobbling back. The New York boy had his arm bones shattered by shrapnel passing through his upper and lower arm near the elbow. He thought his arm was off. Hank was stunned and the hole was filling with blood. Reeling from the concussion, Henry was finally able to run to direct an aide man to the injured man.

The aide man wanted Hank to come along, but Hank was sure he didn't want to go out into the hail of shelling again. As he watched the lone medic go into the woods, he decided to follow—Hank thought it was the least he could do. When they both reached the hole, the young man was pale. Hank helped put a splint on his mauled arm. They weren't sure the man could walk but he would try. They helped him up—one on each side—but he fainted at the edge of the hole. The aide put his arm over the G.I., while Hank braced up his other side. The shelling resumed and Hank wished he had never come back to that foxhole. They both dragged the man across the shell-ploughed ground back to the bunker.

Hank stayed at the bunker and then went out again to find an empty hole. The wounded weren't evacuated until the next morning. Hank hoped the young guy's arm could be saved. He needed his hands for the work he did to support his wife and several children in New York.

Hank made Sergeant that day. The next morning a new Captain came up and wanted to get organized. Platoon Sergeant Hackley recommended Hank for assistant squad sergeant to himself. It snowed and they had remained in position for two or three days. That night Hank, who was in a hole by himself, heard a wounded man calling for help in the dark. On the second morning there, the Captain told Hank to take a patrol through the woods and contact the C.P. (Command Post) of a 22nd Infantry company on their left flank. They were to deliver a message so Hank, with his men, headed out early to escape the German artillery that pounded them every morning.

His three-man squad moved fast, skirting mines and barbed wire along a firebreak trail, going to the right over a hill, down another firebreak, and up a hill to the neighboring C.P. They talked a little there, but Hank wanted to hurry back. The shelling was due. As they crested the top of the hill, the Germans cut loose with twelve rounds of zonal fire on the ridge. The men jumped up and Hank yelled, "Let's go." He was surprised to see that all his men were still on their feet and running toward the pathway. Hank then heard again the wounded man of the night before calling somewhere near them from a mined area. They all wanted to look for him, but it was too late. Just as the barrage started in force, they were able to race through the woods and get to the company position.

Hank reported to the Captain and headed back to his hole. The zonal fire began cutting the trees. That evening the Captain promoted Hank to squad sergeant. His duties included making sure that the men had the supplies they needed. One night he heard a gunshot and went to check it out. One fellow had shot himself in the foot in desperation and on purpose to get out of the war—for a while at least. He had to be taken off the line.

One day Hank found a dead, frozen German NCO and took his magnifying glass, a map reading glass, and about eight bullets for his pistol from him. He didn't take anything else. The next day he passed the dead German again. The frozen man was minus several fingers—someone had cut them off to take his rings. This was fairly common. (In Normandy he remembered seeing a dead German paratrooper with an SS ring but had not taken that either.)

There were some loaded German machine guns laying around and a replacement, a man in his thirties, took one and fired it. Hank raced over to him, face red with anger, and yelled, "What the heck are you doing? Are you nuts? What do you want—artillery poured in on us." The man said he was sorry. He suddenly realized the danger. Hank explained to him never to fire unless he was ordered to or unless he saw a German.

Angrily, Hank tore the gun apart and threw the pieces

away. He kicked at the pin to bend it but only succeeded in tearing a clean cut in his boot. Luckily the metal didn't cut his foot. Two days later, Sergeant Henry Strecker was walking around, checking the guys in the holes and the new replacements. He got into one hole and began talking seriously, just saying that he was thankful that he was alive each morning when he got up. Another older fellow said he felt the same way. Hank left that foxhole none too soon. About fifteen minutes later a mortar barrage started. The older fellow Hank had been talking to was killed in that shelling. He left a wife and several children.

Later Hank and Hackley were with a fellow who kept complaining about his cold feet. Finally, Hackley drawled, "Do you want to put your feet in my pockets?" That ended the griping.

The 4th Infantry Division was being mauled in the Forest. Then one morning, the 8th Infantry Division came in to relieve them. The 4th was being sent to another part of the forest. The men ran out as fast as they could go, slipping and sliding over the pine needle carpet with hardly enough strength left to run. They were only too happy to say "auf wiedersehen" to the place they had called Purple Heart Hill—to the fairy tale forest which had become a nightmare. It was a nice sunny day.

Hank left the Scout knife he opened ration boxes with in his hole, but he had no desire to return for it. They rode in trucks to another sector where they got off and started down a firebreak, with Teller mines sitting unburied along it, and up a steep ridge on a road south of Huertgen where they met the enemy again. German machine gunners forced them to withdraw. In the morning, they advanced and withdrew again to take over the position of another company. Hank couldn't help but notice how happy the men of that unit were to vacate the holes which C Company was taking over. Hank was near a fellow who got hit in the back with shrapnel as he was getting out of his hole. The wounded man gave Hank the new four pocket field jacket he was wearing before he was taken away.

The next morning the outfit got up and crossed the road to a place where some rocket launchers were set. The dirt behind the launchers was all blown away from having been fired. The men dug in. It was wet and soggy that night. The Germans sent three or four rounds over, which fell short into the woods.

Replacements, fresh from the States, an odd contrast to the grizzled riflemen of the forest, arrived to keep some semblance of units in the decimated ranks. Supply trucks brought some wet shelter halves and one of the new men was busy griping about his dirty and wet canvas. Hank was futilely attempting to fix up his foxhole—it was all muddy slop and goo—but he stopped long enough to hear the commotion and got mad at the new fellow. He told him to shut up, fix his shelter with the man who had the other half of the shelter, and quit griping since he might be dead soon and not need a shelter half.

Glen Smalley wanted to be sent back to the aid station

because he had a cold, but Hank wouldn't let him go. With a shelter half and a wool zipper blanket, Hank slept alone in a narrow foxhole.

The next day was Thanksgiving, and a traditional dinner was served, complete with turkey, dressing, coffee, cranberries, potatoes, and peas. It was good. The company stayed put until the next day and were again trucked to a different part of the forest with more open ground and leafless trees, where they dug in for the night. A clever fellow took his shoe off, shot himself in the foot and put his boot back on. He too wanted to get off the battle line and tried to claim that the Germans had shot him but found it impossible to explain how they put a bullet through his foot without nicking his boot.

Hank checked a dead German and took a razor which he needed. Then he wrote a letter home. It was hard to write since he felt low and didn't want his depression to show through in his letter home. Lil and the family worried enough about him. Then he got the sinking feeling that this might be his last letter to them. He gave the letter to the lieutenant for censoring, but the lieutenant said it was too dark to read and that Hank should keep it on him until later. Hank felt there might be no later. He tore the letter up.

They went to the attack and Hank still had a feeling that he would never make it out of the murky, misty Huertgenwald. They passed an old German who was just sitting sullenly in a ravine near some cooking utensils and cutlery. Near some concertina wire in the woods, they encountered more Germans who darted from the firebreak on the other side of the road. No one fired a shot. The GI's reached a Y shaped junction of two firebreaks, where to the left, about half a mile away a German soldier was walking away alone. Hank aimed at him but didn't fire. The German strolled away in peace.

Someone stepped on a mine and was killed. They dug in for the night at the junction. Hank was alone again, feeling terribly sick and dizzy, sure that he would not be able to get up in the morning. He could only picture himself being carried from that hole in the morning. Another man stepped on a mine during the night and was killed.

Much to his surprise, Hank got up feeling fine in the morning. Cautiously, they moved forward again, passing the torn-up B Company position. It was freezing and had snowed. Hank walked past a dead German who had his face pushed in—a tank had run over him. He picked up a pair of German army gloves, green knit wool, which were still tied together and several new, clean white handkerchiefs. Later in the attack, he took the German gloves off, afraid that he might be shot by mistake, and found a pair of GI gloves to replace them.

They were on a hilltop north of Gey late in the afternoon when they attacked again, capturing two young Germans in a woods near the bottom of the hill. They dug in at a creek bed at the end of the day. It was December 7, 1944, and they were above the Cologne plain. Hank stayed with

Parks, an assistant squad sergeant. They smoked and read the Bible. Nothing happened.

On several days, Hank got out and walked around, checking the creek bed. Nothing happened. Finally, a man in one of the holes said he spotted a German in the dry stream bed. One of the GI's had the "back door" of his foxhole facing the stream and could see it easily. He put his gun out through the hole. Hank got ready with his rifle. There was a shot fired, a puff of smoke rose, and the GI yelled. The German had grazed him in the head, and he had to be taken out. The sniper crept away and that afternoon the German shelling began. Two or three men were wounded lightly. In the evening they withdrew.

A platoon from Company A helped carry the wounded men out. Firing was coming from Gey while the company reached the relative safety of another hilltop with a well-built German trenchwork. Hank settled down and was dozing off when Lieutenant Buckmaster came through asking for a man to go down and lead some other fellows up from the foot of the hill. Parks piped up that he didn't know the way but that, "Strecker does." Hank was almost asleep when the Lieutenant came up and announced, "Strecker, Parks says you know the way down. How about showing some of the Company boys where to go and get some of those guys out of there."

Exhausted, Hank was upset with Parks but grabbed his rifle and started down the hill, eyes straining in the dusk, expecting machine gun fire at any minute. He made it down and found all the men sleeping. He was afraid that if he scared them he might get shot by accident, but luckily once again—nothing happened.

These were Lieutenant Shasteen's men and Hank started to lead them out. They met some men carrying a very heavy-set wounded German. They kept urging the other men to "keep going, this guy's awful heavy." They hurried. Hank asked where everyone was, but the men kept moving and replied only that the end of the line should be coming up soon. They met Lieutenant Shasteen. He asked about his men—some were missing—and he angrily told Hank to, "Get back down there and bring all of them up out of there."

Hank went back alone and found more men at the bottom of the hill. He ordered them, "Let's get going" but they just moaned and said they were too tired to move. Adding to the problem, they had wounded men waiting on stretchers and Hank countered, "OK, let's have four volunteers to carry each of these litters." No one stepped forward. Hank then assigned men to carry the wounded and was really getting angry.

After trudging for a while, the litter carriers were exhausted but no one offered to change places with them, so Hank had to order a rotation. The soldiers started beefing that the wounded were, "too heavy and we can't carry them anymore." Hank had about all he could take and spit back, "Listen, this war's not over YET and someday you might be begging someone to carry you." Finally, jeeps picked the men up.

They started to dig in the following morning. It was rocky ground. Parks lay in his sleeping bag smoking complacently as the shelling resumed. He couldn't get out of his bag when the zipper stuck tight and made quite the scene as he struggled along trying to free himself. Hank dived into a hole and then moved farther up to one near a fire watch tower as the shelling continued. He got sick. Later he crawled into a deep long covered hole with three other men.

During the day, they had to go out and take water from the canteens of the dead laying around them. One Graves Registration man came along picking up the dead when the shelling resumed. One fellow followed by another tumbled into their well-covered hole and the startled occupants yelled jokingly, "What the heck are you guys doing!" The visitors left when the barrage stopped. They managed to share a breakfast box ration with eggs, coffee, and fruit packages, which were their favorites.

Rations came packed in wooden crates with twenty-four boxed meals of one type in a crate. The meal was packed in a rectangular, wax-coated cardboard box. On one side of the box in plain black letters was printed Breakfast, Supper, or Dinner. A breakfast K-ration contained a cylindrical can of chopped ham and eggs (mostly eggs) or bacon, a foil of Nescafe coffee, a fruit bar, three lumps of sugar, and four cigarettes, (Old Gold or Camels.)

The supper ration held four cigarettes, potted corned beef with apple and carrot chips in a tin, a foil of bouillon soup powder, a chocolate bar (the first ones were semi sweet but later in the war they were made sweeter and lighter — hard and about ¼ inch thick (the famous "Shokolade" begged for by German children), one stick of chewing gum and eight Nabisco crackers (thick, hard, and brown with a sour aftertaste) — later made more like soda crackers and finally some toilet tissue. The potted meat was mostly grease and was thrown out by the men of Company C, until Hank discovered that by folding the tin lid back over their coffee cups and holding it over a fire it tasted great when cooked until it fried. When he discovered this, a whole window sill pyramid of the discarded cans disappeared one night in Germany as the men fried them up.

The dinner was a bulky affair in a container like a peanut can. There were three varieties of supper, a vegetable stew, a greasy hash, and meat with beans (the taste favorite). Hank was especially good at improvising food improvements. He would make a lemon pudding by putting his lemon extract in water, crumbling crackers into the mixture, stirring, and heating. For chocolate pudding, he grated up a chocolate bar, added water, stirred, and heated. The bouillon soup was better with crackers added. The wax box container could be burned to heat the ration and didn't create too much smoke. But the GI opinion was, "with food like this, who need's enemies."

* * * * *

Finally, they were relieved by another outfit and headed out to the southwest around Huertgen and to the north again. On December 12, 1944, Hank sent a Christmas card home. It was a simple card showing a cartooned GI sitting on top of the Eiffel Tower above noted Parisian landmarks and holding a Stars and Stripes banner reading "Merry Xmas."

On the back Hank wrote:

"Dear Dad and Kids: How are all of you? Hope you are all feeling well. I'm fine. This is the best I could do in the line of cards, Kids, so I hope you don't mind. I know you won't, you know the feeling that goes with this card, Kids, that's all that matters. I sure wish I could be with you this Christmas but that's impossible, so we'll have to wait until next year. I'll be thinking of you and pretend I'm with you. I wish all of you a very Merry Christmas and a Happy New Year. Take care, Kids and Dad, I'll try to write you a letter later on.

Your Big Brother,

Henry V …

* * * * *

The Regiment moved from Schevenhutte, Germany to Junglinster, Luxembourg to relieve the 329th Regiment of the 83rd Infantry Division on a twelve-mile front. They were trucked in for a rest in this quiet sector and did not know how long they would stay there. Rumor had it at several weeks or a month. After one night in Junglinster, Hank was in a second group given a forty-eight hour pass to Paris. He cleaned up with a lukewarm bath in a cold barn with large cracks in it. Trucks took them to a schoolhouse outside of Rheims where they stayed overnight.

In Paris, they had dinner at a hotel dining hall and Hank, with two other fellows, set out to see Paris. They walked the streets, visited several nightclubs, and had some drinks. That night Hank took a bath in their hotel suite and the following day went out again with his buddies. They had their pictures taken and the French photographer assured them that he would have the photos mailed to them. Two girls came along, and Hank was promptly deserted by his two buddies. He didn't mind and spent the time walking and rode the subway until 12:00 when it stopped running for the night—something he had not planned on happening.

He was let out at a five way intersection and wandered lost past the huge dark-looming building of the Louvre and down the Champs-Elysee to the Arc. He tried in vain to get his bearings by the stars but could not locate the

North Star that night in Paris. A Navy man was crossing the street when a speeding cab almost ran him down.
Hank walked past the Arc d'Triomphe and on out over the
Seine where he found a subway map of the city. There was
a marker lamp illuminating some torn up pavement, so he
loosened it and drug it over to the map, tried to raise it up
to see the map but was unable to and had to put the heavy
lamp back.

He wandered past a Red Cross aid station and French-
men standing by their bicycle cabs chatting in the late
night. In desperation, he asked them for directions. They
rattled on and gesticulated wildly. He wished more than
ever that he could understand them. He walked until six
in the morning when he found the hotel. The other fellows
were going out to see more of Paris. Hank was going to
bed. Later, he went out walking again and met an elderly
man on a street corner who spoke English. Oddly enough,
he told Henry that he had been an American doughboy
in World War I, had met a French girl, married her, and
stayed in Paris. The following morning trucks came to ride
them back to Rheims for the night. Hank went to a night
club there and then bedded down in the schoolhouse.

At daybreak they rode back to Junglinster and Hank,
happy and eager to tell everyone about his trip to the City of
Light and enjoy a good quiet rest, entered the house where
he was quartered. Inside everyone was rushing around
frantically packing their equipment and the fireplace was
already out. Hank was startled and demanded, "What the

heck's wrong!" He hadn't heard the news. Someone answered, "The Germans are coming, that's what." Hank got boiling mad—his dreams of rest were shattered—and the war wouldn't be over by Christmas.

The Germans had begun their last large scale counter-offensive, Operation Christrose, Watch on the Rhine. ***To history it would be known as the Battle of the Bulge.*** The offensive was a gamble that Hitler insisted upon, and von Rundstedt deplored. German hopes were high for an attack that had to be bold and quick.

* * * * *

Within twenty-four hours the Germans were thirty miles into the American lines on a sixty mile front. Panzer divisions were racing for the Meuse and Liege Rivers. The 4th Infantry Division was in the southern sector at the line of Berdorf, Echternach, and Dickweiler in rolling country with good roads. In the south, Rundstedt planned to break through at Echternach, head through Luxembourg, and drive west just as in 1940. Brandenberger's infantry troops, attacking in that sector, could bring final victory.

Hank's outfit marched for about two hours in the snow that night to a village near the German lines. Hank went into a house and tried to sleep on one side of a room, but every ten or fifteen minutes the Germans lobbed shells in, so the GI's finally scrambled to the basement and took shelter near a wall—but the basement was flooded with about five inches of water. They were anxious to get out of that, so they went back upstairs.

Three or four guys invited Hank to get on a bed in the house with them to stay warm. They could talk since sleeping was impossible with the shelling going on. In the morning, they moved into the battle. The Germans were breaking through the woods between the little towns and chopping up the American lines. The 212th Volksgrenadier Division began attacking them at 0945 on 16 December 1944. On 17 December, the outfit started to dig in near a clump of trees along a road until they were ordered further up the road to a new position.

Hank went on patrol, passing another road, some pines, and a beech woods to check for Germans. Two Germans were spotted crossing a field and heading into the pine woods, but no one fired a shot. The patrol circled down across a dry creek bed and up its side. Then one of the GI's said he saw a German "up there" behind some trees. Hackley fired. A tall fellow in his forties was griping that he had come over to kill Germans and hadn't done anything yet. He asked Hank, "Strecker, I see a Kraut up there, should I shoot?"

Hank asked if he was sure it was a German. If it was, he could go ahead and shoot. Hank couldn't see anyone from his position. The man fired at the German who was standing framed in the fork of two tree branches and got him between the eyes. They took the German's Luger and captured two teen-aged German soldiers. One, wearing dark round framed glasses, was carrying a sniper rifle with a tele-scopic sight. He was grinning from ear to ear and appeared to be very frightened. Snipers were not well received.

Sometimes the Germans knew which outfits they were fighting, and several had recognized the Ivy patch on being captured. They knew the 4th was a tough outfit and they did not much like having to go up against what they called "the terrible green cross."

As a defensive post, the men moved toward the left and stayed there that day and part of the next. Lieutenant Buckmaster was hit so Hank and Hackley took over the platoon for a time. The Germans had captured one pla-toon in the fast moving battle and later Hank found out what happened to them. The Germans were checking their new prisoners when one of them found a P-38 clip on an American. A young German wanted to shoot the prisoner but before he could fire, a German officer came up, pushed the younger soldiers' rifle barrel down and told him there would be none of that. Just then US tanks began to rumble up through the woods and the alert GI's talked the older German officer into surrendering with his men. They knew the war was over for them.

Hank also learned that Lieutenant Shasteen and most of his men were found dead in their foxholes. On the 17th, the Germans had overrun them and apparently, they had no chance of surviving.

The outfit stayed overnight in position along a firebreak. The Germans were pushing through the lines above Echternach and that afternoon Hank's group was trucked to Consdorf under shell fire to hold the Germans off. They were moving fast and saw engineers trying to fight their way out. The village was captured and the GI's advanced through a pasture and ravine while the Germans shelled the town.

At dusk, the shelling stopped and in the evening, they went to the outskirts of Echternach. There they took up a defensive position on a pine crested slope above the road. The snowfall that night left a blue haze wafting around the trees as Hank peered down the slope to his right along the line of the woods. As he stared, he could hardly believe what he saw at dawn. Germans seemed to be popping up out of the snow all over the field. They were calmly and surely getting up from their holes and into a V formation for their attack. The GI's were worried. The Germans were walking steadily toward the edge of the pines and the woods.

The Company called for support. A lieutenant came and gave the range for the artillery to open up while two light tanks clattered up the road. The Germans pressed boldly across an open field and moved forward undaunted. Company A let loose with machine guns and the valley became red with explosions and black with smoke. The tanks pulled into the open and began firing so fast that their cannon barrels softened and melted. They didn't stop until their overheated barrels were spinning and writhing, spewing shells wildly in all directions. The German assault on the hill was futile and the fighting stopped. A hundred and fifty-four Volksgrenadiers lay dead in the snow. Only three survived the attack and they surrendered. Hank found one dead German who had been lugging a .90 mm. mortar and shells weighing over a hundred pounds by himself.

The 212th Volksgrenadiers were unable to break through at Echternach. After six days, they were decimated by the 12th Infantry Regiment. On December 21, the Volksgrenadier's 212th's, 423rd Regimental surgeon, a captain, was captured and said, "The terrible price in dead and wounded that the 212th Division has paid for the meager success achieved is exorbitant."

It was December 22nd when the last German drive against the 12th Infantry Regiment was over. On December 24th the outfit was relieved by the 5th Infantry Division and Hank was glad to get a Christmas rest. The men hurried down the road at a fast pace, even though they were exhausted. They turned right to Bourglinster, Luxemburg and were all billeted in its vicinity and at Berdorf for Christmas and three days as division reserve. The house where Hank stayed had a small Christmas tree in the living room, a fire, and some Christmas cookies. He and five other men sat in the living room talking and singing Christmas carols until it was time to sleep — on the floor.

Christmas breakfast was fried eggs, bread, and coffee. Between ten and eleven there was mail call. For Hank on this surreal Christmas day, there were twelve packages and letters. Lil had gotten nearly every one of the Strecker aunts and uncles to send him something—cookies and hard candies. Hank put the candy in a bowl on the table and passed the cookies around. In the afternoon, they were invited to a Catholic church for caroling. The people sang *Adeste Fidelis*, *Stille Nacht*, and another Christmas song.

Henry had time to write a letter home to his father, sisters Lillian and Ellen, and brothers, Edward, and Alan. Here is the letter:

Dear Dad and Kids: *December 25, 1944*

Hope all of you are feeling well and happy. I'm feeling fine. Today is Christmas, hope all of you had a nice one. You won't hardly be able to believe this because I could hardly believe it myself. Last night I was laying in my foxhole freezing, with shells blasting around one minute and the next hour or so I was in a small town. Then in a nice warm room, in the home of some family. In one corner was a Christmas tree with lights. I could hardly believe my eyes. We sang Silent Night and had a swell time. It seemed like we were in a different world.

Then, to top things off today, I received twelve Christmas boxes. Your old brother certainly is a popular guy back home. The boxes were from you kids, Uncle Al and Aunt Kate, Uncle Otto and Aunt Ida, Aunt Hilda, Uncle Carl and Aunt Marie, Aunt Kitty and Uncle Walter, Betty Skimmerton, Ernie Hanfbauer, and Bea. We ate so much sweet stuff we could hardly eat our nice turkey dinner. This afternoon we went to church, a Catholic choir sang some carols for us. We also sang some.

What more could a Joe ask for besides being with sisters, brothers, and Dad. It was all wonderful. I thought of you kids often throughout the night and day and just about what you were doing at certain times, also of the things I was doing last year at this time. I hope I can be with you next year at this time. I thanked God that He picked me along with many other boys to have such a nice Christmas in a war- torn country. Said a prayer for the boys up there holding the line too. May God soon end this war.

I went to Paris awhile back, well just about two weeks ago. Bought you, Lil and El, a compact. I hope you get them before too long. They aren't on their way yet, but I think they will be shortly. I went out with some French girl. Had a very nice time. I didn't get liquored up either. Don't swear anymore either. When a guy comes through what I did, he realizes the good Lord is really someone to be worshiped and the sayings of the Bible are something to follow.

About that close call I had, Kids. I was walking through the woods and all of a sudden, Wham! A mortar shell burst in the top of a tree I was under. I was showered with snow and black dust. I kept going to get away from that spot. I glanced down the front of me and saw two shrapnel holes in my gas mask strap. I examined myself closer and found that the shrapnel had gone through my gas mask strap, raincoat, fatigue jacket, field jacket, cut a rip in my pistol holster, glanced off my little pistol butt, after cutting a gash in the steel part of it and came back out the way it went in. From there it went through my gas mask carrier, through an interpretation book, and laid in my gas mask carrier at the bottom. Slightly close What!

My little pistol saved me without me having to fire it. The shrapnel went right through the middle of my fatigue jacket pocket. My Bible would probably have stopped it if my pistol wouldn't have been hanging there. I carried that gun from Cherbourg, never took it off except when I had a chance to wash. God was with me, is all I can say. I'm sending home the piece of cloth from my fatigue jacket pocket. Keep it for me, will you?

By the way, the gun shoots as good as ever. I hated to see the gun get marred but it's not too bad and after all, it probably saved my life. The clock just struck twelve. One more Christmas gone by. A very nice one except that I wasn't with you Kids. I hope by next Christmas I will be. You Kids don't know how much your package meant to me. Everything was delicious and I want to thank you from the bottom of my heart. Tell my little buddy Alan to keep praying for his brother and all you Kids do the same. Take care Kids and until the next time as always,

Your big brother Henry V …

* * * * *

For service in the Battle of the Bulge, the 12th Infantry Regiment was awarded a Presidential Unit Citation. The photos taken in Paris arrived, but Hank never received his — he received one of the other fellows and was able to give it to that man. He did receive a pencil sketch, done in Paris, along with a photo of the artist. The caricature artist was Dessin Bornett who sent his address on a photo of himself.

At New Years, the 12th Infantry Regiment was on a defensive line from Bollendorf Pont to Echternach, facing the Siegfried Line. All was quiet in their sector that first night of 1945 until someone fired a machine gun burst. The men were above the Sauer River and could hear the Germans talking, working, and digging — using a small motor across the river.

At one point, Hank had to meet three replacements at the company outpost foxhole. As he went to leave with them, a mortar shell came in, and he had to dive back into the hole. Luckily it was a dud, and the new guys had a good nervous laugh. Later he took them to change the guard. One of the guards whispered "Halt" loudly as the bug-eyed replacements kept moving. Hank heard the guard's rifle click and quickly exchanged the password for the night "Fire" with the reply "Engine" from the guard. The guards were startled by the replacements noisy approach and Hank was also worried because he had to remind them to be more careful and keep their eyes open.

Their next stop was at what appeared to be a film stars house along the Luxembourg road nestled in a pine forest, which would be used as the C.P. The house had full-length mirrors on the walls and one room had large studio lights set up in it with movie star photos strewn on the floor. On a knoll down from the house sat a pretty little chapel and in the back was a garage which could have held eight or nine cars. (This was the Chateau de Lauterborn, which was originally a farmhouse leased by the Benedictine Abbey in 903, then was renovated many times, finally as a Baroque style chateau in 1784. In 1797, it was sold by the French Republic and had over a dozen owners over the years until in 1936 it was leased by a Baroness to a Luxembourg film star who was used by the Nazis.)

The Germans had left it crammed full of bicycles. The GI's were suspicious and moving carefully found a small panzerfaust with dynamite rigged to it. The charge would have blown the place up if someone had just tried to jerk one of the bikes loose. In the house, the GI's had a blaze going in the fireplace and pulled a couch up to it so they could warm their feet. Their backs froze anyway. They took two of the bikes out and enjoyed riding them down the smooth sloping lane to their outpost. Pumping back up hill was another matter. Hank spent some time at the outpost and wrote about what happened there:

"My men and I were stationed on an outpost about a mile and a half from Echternach. We were staying in the cellar of a house on the road leading into Echternach. The town was visible from our position as it lay far off down the road in a valley. The Sauer River ran through the middle of the town. On the second morning there, I got up and went outside. There was a barn behind the house, and I decided to have a look around. While I walked around behind the barn, I noticed patches of brown grass sticking up through the snow. The sun was shining brightly on the snow and the air was clear and fresh. It reminded me of some of the wonderful winter mornings back home when I would take my gun and hunt for rabbits.

"I started checking some of the patches of grass when to my surprise in one of the patches sat a large Belgian hare. I couldn't fire a gun for fear of broadcasting our position to the Germans. I thought maybe I could get him with a club, but he jumped out immediately, ran around the barn and through a doorway into the barn. I was in hot pursuit. I closed the barn door, and the rabbit was mine.

"After dressing the rabbit, I went up into the kitchen of the house and found a frying pan, a pot, flour, and a can of GI butter that some of our tank boys had left. I needed some salt and pepper but couldn't find any. Then I happened to think about the bouillon powder we used to throw away out of our supper rations box. It had salt, pepper, and a few other spices in it. I gathered up my utensils and headed for the cellar. I found a few packages of bouillon powder and I was all ready to play Henri the Chef. There was a stove in the cellar and I had a good hot wood fire going in minutes.

"As I sit here and write this, I can't remember if the

smoke was going out the chimney on top of the house or if I had the smoke going out the cellar window. At that time, I never even gave it a thought. If the Germans did see it, they never did anything about it. I sprinkled bouillon powder on the meat first, then dipped it in the flour. The GI butter in the pan was all set to receive the rabbit. I fried the rabbit first and then simmered it in the pot. By that time, our mouths were watering, and the aroma was extremely appetizing. We devoured the rabbit to the last bit—thanks to my experience in cooking while in scouting and from watching my mother cook.

"The next afternoon I decided to get nosey and went down into the outskirts of Echternach with one of the boys. We inspected some of the houses. In the last house I went up the steps to the second floor. As I stepped up on the landing with my pistol in hand, something caught the corner of my eye. I whirled around and almost froze in my tracks. There I was standing face to face with someone, gun pointing right at me. It took me a few seconds to realize it was myself in a full-length mirror.

"I hurriedly looked through the rooms. Clothing was strewn about, drawers were all open, everything was a mess. After we left the house, we were undecided as to whether we should go on into town further, for there was a huge champagne factory down there. We thought maybe we could scrounge up some champagne. Thinking that there might be some Germans down there, we changed our minds quickly for there was just the two of us.

"As we arrived back at the outpost, two P-51 Mustangs from our air force dive bombed the champagne factory with some five-hundred pound bombs plus a few other large buildings down in town. We were glad we decided not to go down there."

* * * * *

One night they moved from the outpost up a hill and into a valley, up another snowy hill to relieve a different company for a week or two. They were still along the Sauer near Echternach. The Regiment was put in Patton's VIIth Corps and all men were ordered to shave their faces clean. He was caught going from the CP to his post and reminded about the order to shave. Reluctantly, Hank parted with his signature goatee and mustache.

Hank went walking by himself toward the woods and along a wire fence where the Germans had been entrenched. German rifles and equipment littered the ground. A dead German lay there face down. Hank turned him over, pulled his Schmeisser apart, took a 39 Cross of Iron, some other medals, and an octagon shaped, bone stemmed pipe.

They stayed a week in the snow. Hank was in a foxhole with one other fellow — covered by a raincoat. They were so crammed together that at night they had to turn over together, but it helped them stay warm. K-rations were the only food available.

On January 18th, the outfit began their attack again.

They were spread out on the Luxembourg line facing the Siegfried Line across the Sauer from Bollendorf Pont to Echternach. The Second Battalion, attached to the 8th Infantry Regiment, crossed the Sauer near Bettendorf to attack the German bulge on the south flank. The Captain told Hank their assault would be on Vianden. Company C was put in reserve and watched the rest of the Regiment cross the river, make the attack, and then crossed the river themselves. One of their men lay dead on the road—his eye hanging out on his cheek. It was very de-moralizing.

Company C was to check out the town and the area beyond. They could hear a German machine gun in a house up ahead and the Germans managed to throw rounds of artillery in on them. Lieutenant Pochinchuk was getting shook up. Darkness kept them in the town. One man had his machine gun set up in the door of a barn and when he heard someone prowling around, he pulled out his .45. When the prowler came into sight, he shot him in the shoulder. The man let out a blood curdling scream. It was one of his own men.

The GI's were sleeping on wooden benches inside when more shells came in, knocking out the lights—and the men onto the floor. They improvised, sticking some old socks in bottles as wicks and lighting them. The Battalion was told to get a good sleep. They were going to make a night attack. It was nearly impossible to rest. At 11:00 that night they set out, advancing slowly. The machine gun nest located up the road worried them, but they did not encounter it.

A hill was crossed and then they went into a valley where they stopped. The Captain sent out a patrol to see what lay ahead and the scouts returned baffled. Vianden was nowhere in sight. The Captain ordered attack formation and the men went up over the hill. They found the town.

Hank looked down the long, silent line of men. They were darkly defined masses in the white field of snow. Tensely they waited for German guns to open up, but none ever did. All the houses in the town were checked. The inhabitants were there but only one German soldier was found. He had been sleeping in a hayloft.

The Company slept and ate breakfast, remaining until the afternoon, when they started a night time march to another town which they occupied. The following day, they spent cleaning their equipment and resting. On the next day they were ordered to move out and Hank began rolling his blanket up on the floor. When he stood up, his helmet crashed into a round, glass-globe light cover that hung directly above him. The bottom of it tinkled neatly to the floor. Hank felt terrible since he was trying not to ruin anyone's property. It seemed ironic — there was too much destruction as it was. Two old, pipe-smoking German men lived in the house and Hank forgot his tobacco pouch on the kitchen table. Well, that seemed fair. The old men could probably use it.

In a little schoolhouse in one of the towns, they sat around cleaning their weapons and magazines. Smalley was cleaning his BAR, put his magazine in and had the rifle cocked when he accidently pulled the trigger. Three rounds shattered the window frame. Everyone was startled, including Smalley who sat with his mouth wide open, all at once, frightened, and amazed. Amazed too that no one was hurt.

The following day, the outfit rolled once again through Bastogne in trucks, on the heels of the Germans. The town was now bombed to rubble. Jeeps full of wounded were coming off the line as they moved forward. They reached a pillbox and dug in. Sergeant Whitey (he had white hair and always carried a sub-machine gun) joined Hank. Some Screaming Meemies fell in. Since Hank was a platoon guide, he got into a shack with Lieutenant Pochinchuk. Sergeant Mueth passed some whiskey from home around and the Lieutenant shared some of his whiskey ration. Water seeped through holes in the shack making it rough.

Two days later they moved onto the line about seventy miles north in the Schnee Eifel, toward the Siegfried Line. This would be their second time through the same place. Hank walked down a strangely familiar road and then noticed a statue of Mary in a wood shrine off the road. It was the same one he had passed several months earlier. It was below freezing, and the snow was knee deep. The Germans had pulled back the bulk of their troops once again and left only small rear-guard units with a little artillery. They were saving their troops for another day. The Volksgrenadiers were sacrificed.

Hank passed dead cows with stiff, swollen legs. The

roads were torn up from shells and the trees were chopped and riddled. To the left, an .88 shell came in so close that Hank could still hear its whistle ringing in his ears after it exploded. The Battalion under Lt. Colonel Golden was to assault Elcherath. Aerial bursts exploded in the town as the engineers struggled to repair a bridge leading to it. Strung out along the road, the men started into the town and had moved past a house and barn when twelve rounds of .88's hurtled in on them. Hank hit the ground and landed heavily on the bulky field glasses strung around his neck. He jumped up and dashed into an old house. Joe was helping the men who were hit along the lane. Later Joe came in and called Hank yellow.

Tempers flared and Hank countered that he wasn't yellow — just scared. And in combat, sometimes everyone was scared, and they all knew that. The next day the men moved up and were replaced by another part of the company. They stayed in a barn as part of the reserve. Hank wrote home.

Front line veterans were needed for Officer Candidate School and Lt. Pochinchuk recommended that Hank be sent to it. The Lieutenant gave Hank a pair of gold shoulder bars and Hank proudly put them in his pocket. He was on outpost duty a couple of nights later and the second time that they were coming in it was pouring rain. Hank started to joke about it, "I don't see why these guys have to sit out here getting soaked and sick with pneumonia. I'll have to see the company commander about this ... " They could still laugh. They had to.

Elcherath was the first town on German land captured since December 16th. It was the fifth time since September 13th that the 12th Infantry Regiment had been across the German border. Lt. Pochinchuk left at this time. He needed help and a rest. Combat had been shaking him up for a long time and he was getting more anxious. One evening Hank was with him as they went up a hill. An American sergeant lay there dead with his head blown off. That was when the Lieutenant told Henry he needed to leave. It was too bad. He had been a very committed, good officer.

Trucks took them to Brandscheid. The Germans were gone, and the little burg was a filthy shambles. The roof of one house was burned off. They took up a defensive position.

Captain Larcade wanted to see Hank and sent him to see a Major in a pillbox office for an interview about OCS school. Hank was eager to get the training, just to be off the line for a while and perhaps live longer. The Major played with his pencil, turning it end over end, as he fired questions at Henry. He was asked if he was willing to fight Japan when the war in Europe was over and he said yes. He was asked about several articles which had appeared in *Stars & Stripes* and he answered them correctly since he did read the army paper. There were more questions about different outfits and their positions, which he also answered correctly. He filled out an application and several days later took a physical at another house. The doctor and an aide made him do some jumping exercises along with some other tests that about knocked him for a loop. He was nearly exhausted and turned in his heavy BAR.

They left Brandscheid in trucks. Hank clambered onto the truck and was asked to man the .50 caliber machine gun mounted on it. He had to sit higher and was almost knocked out by the machine gun being swept under tree branches as the truck sped along the road. Low hanging branches continued to whip Hank in the face, but he held the gun firm. Luckily, the Luftwaffe did not appear over the lone convoy.

The First Battalion went in and took Ihren and Schweiler. The next move would be on Prum. The town was unforgettably bombed into rubble. Along the road, Hank passed some huge German artillery shells and Germans on a steep hill sent 88's onto their column.

One snowy night in January 1945, they trudged into a little town, crossed the Our River on a pontoon bridge, and marched an hour longer before settling into the cellar part of a barn for the night. It was small with a hard floor. The men found some German bread and tried to eat it, but it was tough and left a sour aftertaste. The day was spent just sitting around and a lot of kidding went on with so much extra time. One fellow was acting like a radio commentator while the others laughed and enjoyed the show. Later they moved up farther again, cut off onto another road and up toward some beech woods and dug in along the backside of a hill. Over the hill was a cemetery and a small town. They stayed that day and snow camouflage was given out. It consisted of big white overalls and hoods for the patrols.

A night later, they moved into the little village which the Germans had abandoned and went into the cellars. A lone German medic walked up the road. He surrendered and went to the rear with some GI's. Headquarters sent word that rotation furloughs were being given out. Hackley got one and left that night. All the guys were happy for him. The next day they started up the road through the cemetery. In one corner of it sat an abandoned German .88 on wheels. As Hank passed it, he felt good. It was another sign that the war was almost over and that was one more gun the Germans wouldn't be firing on them again.

Stars and Stripes published a booklet in Paris at this time under the title *Famous Fourth: The Story of the Fourth Infantry Division* with information provided by Major General H. W. Blakeley and his staff.

Quote from Lone Sentry, H.W. Blakely Major General, Commanding and staff (for Stars and Stripes) 1944-1945.

"Scaling the Schnee Eifel in a snow storm, the 8th closed in on pillboxes and entrenchments from the rear to recover in two days all that segment of the Siegfried Line which it had won in September. The 22nd took the fortified town of Brandscheid, which previously had withstood all attacks. Double Deucers drove through snow, rain, and mud, deeper and deeper into the Rhineland. Germans fought and fell back from village to village: nowhere did they stand more than a day. On Feb. 9, the 8th crossed the

Prum River. Two days later, the 22nd took Prum. Pausing long enough for other divisions to draw abreast, the 4th, along with the 11th Armored Div., pushed on to cross the Kyll river at the beginning of March. A task force under Brig. Gen. Rodwell made a dramatic twenty four hour dash which carried it more than 20 miles, capturing Adenau and Reifferscheid. The division took added pride in turning Adenau over to Maj. Gen. Troy Middleton, VII Corps Commander. Gen. Middleton was an officer in the old 4th which had occupied the town 27 years earlier. As General Rodwell's force was fighting forward, orders arrived to move the division 200 miles south, to Gen. Patch's Seventh Army. New problems, new battles await the 4th, but it faces them with calm, certain confidence that it will do what it always has done — accomplish its mission."

* * * * *

From February 11, 1945 to February 26, they were on the west bank of the Prum river in a defensive position. Hank found a .32 pistol like the one he had and gave it to another man, Thompson. They took the hill and stayed in one of two large empty square, concrete reservoirs there. A day later they attacked and took more high ground. Tanks advanced only to be knocked out by a German TD, but the infantry moved up to some houses and beyond.

The next stop was Rommersheim where houses were in flames. As the men went into the German town, through a pine woods and over a field, two Russian boys yelling "Me Ruski, Me Ruski…" came up to them wanting to surrender and be protected. In one house, Hank found a German officers coat and dress shirt with SS insignia---the Germans apparently had left in a hurry. Hank took the brown leather belt which was lying on the bed. It was one of the finest he had ever seen, of good thick leather and the notch marks in the belt clearly indicated that the owner had been none too thin.

The GI's checked the basement with their rifles ready, calling out "Kommen Sie hier." Two German soldiers came out—just kids. They surrendered and left their home addresses with the family in the house so that their relatives could be notified that they were captured. Hank found a harmonica in one house and couldn't resist playing it a little before leaving. The outfit stayed several hours and late in the afternoon moved on up the unending road. Abandoned German field pieces lined the highway. Some of them had large wagon wheels and iron rims. The Germans shelled the farther part of the town as the men moved from the village into a woods, across a creek from another hill and road.

On the hill was a house sitting above a railroad line. Field glasses were called for and the men took a long, good look at the terrain. Germans were dug in a few feet from the house and smoke curled from the chimney. Captain Larcade appeared and said, "Well…I guess we better try it." Hank quickly saw the disadvantage and spoke up,

"Captain, look at that railroad bed … there's no cover up on the road either. We'll get torn up there."

The Captain countered, "We've gotta' attack. Battalion says we gotta' move up." Finally, the Captain could see the futility of an attack and held the position. There were some huge rocks nearby in the woods and that night Hank could hear some whistling. It may have been birds, or it may have been the Germans signaling one another. At any rate, the next day the Germans had vanished, and the Americans moved to the next town — Mullenborn.

They were assigned to houses while the German occupants were told to evacuate. Captured wine which had been held for a while was given out along with cognac brought up in a jeep. Hank didn't feel like drinking, but he did sample a few different wines. Other fellows started to get barreled up and the show was on. Two men started to fight, and one fell or got pushed through a window, breaking it. Sergeant Joe Juartez stormed in and told them if they couldn't hold their liquor, not to drink it.

He was half pie-eyed himself. Later Hank heard some loud arguing and went into the next room to find Joe arguing with another man. Hank warned Joe that he had just heard him tell two other fellows, "If you can't hold your liquor, don't drink it." Joe settled down.

The house had a large stone fireplace in the middle of it with a chimney running up through it, built like an enormous smokehouse. On the second floor, an iron door opened into the chimney and around the inside hung several hams and slabs of bacon. The GI's took them downstairs and started a ham fry. This German house was going to be cleaned out. There was a jar of cream on a shelf and one on the stove warming. They found a churn and decided to make butter.

Hank watched as the eager group churned away madly with no results. A Georgia boy suggested adding a little warm water since the room was pretty cold. That did the trick. The butter was being cleaned up along with the salty ham, when Hank noticed a large oval vat sitting on the floor with a wooden lid on top. Curious, he lifted the lid. A terrible odor arose and almost knocked him over. It made him drop the lid quickly. Whatever it was (maybe sauerkraut fermenting) they would not disturb it again.

Meanwhile, several fellows had raided the German family's closet and one had taken a beautiful gold watch. Sadly, the German family wandered back in and left speechless. In one room, squad leader Joe Blackwell of Marshall, N.C., was drunk, waving his .45 around and bellowing that no one better insult his mother, or he would shoot them. He was getting louder when Hank walked in to break it up. He stormed "Look, Joe, knock it off. I wouldn't want anyone to insult my mother either, but no one here said anything against your mother so shut up. I have a mother too and she's dead."

Joe quieted down and never said much after that. Hank remembered him as quite the individualist and after a few days in a foxhole, with his reddish beard stubble growing, he looked like a prairie dog when he popped his head up to look around.

Hank went into the next room and sat facing the door with his .32 ready, just in case Joe got mad and came in looking for trouble. Joe Balsam of Pennsylvania, who had a trucking business was another independent — and drunk. As they packed up to move, he had tied a ham to his belt. Hank was glad he wasn't drunk since they were going into action. Hank watched the ham swinging back and forth at Joe's side. It was pretty funny. They went through another town and saw bushels of salty hams left behind by the Germans.

Outside the town in a field lay two dead GI's. One had a box camera and Hank picked it up, carried it for a while and then dropped it. He often wondered if there was film in it and why he had picked it up to begin with.

Machine gun and rifle fire opened up on them as they settled into a woods. Joe promptly jammed his rifle into the mud. He wanted to fire it and blow the mud out, but Hank wouldn't let him, fearing it might ruin the barrel or injure them. Instead, they cut a sapling to clean the barrel and settled down. Joe still carried the ham, but now it was a filthy dirty mess from being dragged along through the mud. They took some German prisoners and other Germans began to fire on their own men as they surrendered.

Roth had been taken, a river near Dohn was crossed, and the outfit stopped at Walsdorf.

* * * * *

The Rhine lay ahead of them. It was spring as they entered Alsace on trucks and went through a heavily shelled town, Haguenau. Hank stared at the shell-hole torn roadside and blown apart trees as they rolled along in trucks. They were going to another attack and stopped late in the afternoon. Plans were changed and they were ordered back to Haguenau, where they found a bar. It was empty. Some company wits quickly found a wine factory. Several men rolled a twenty-five pound barrel of clear spirits down the road to a barn. There they heaved it onto a table and put a spicket in it. The brew was drawn into a spoon and lit. It burned. Then some spilled into the straw and the men had to frantically beat out the fire before it spread.

Finally, one day, Hank decided to try some and put it in a bottle, to see how much he could drink. He took it back to the house where he was billeted and wrote a letter home. He would sip and write for about fifteen or twenty minutes before he had to quit. Later he put a little wine in the alcohol just to cut it. He didn't want it to go to waste but it was too strong to drink straight. There was a lot of Five Star Hennessy around too. It was a brown whiskey and Hank tried a taste of it. By this time, he could drink anything Europe had to offer.

Early in the spring, Henry remembered stopping in Rambervillers-Aux-Bois where their helmets were painted, and the kitchen was set up for regular hot meals. Hank was quartered in a house with a woman who had two or three daughters, a cow, and a son who was a policeman in a neighboring town. With some French and German coins in his hand, Hank asked the woman to sell him some "Eier fur Geld" (eggs for money.)

She acted like she couldn't understand and then said no, she would not sell. Hank took several short walks but did not get to see much of the town. The highlight of the town was a forty-year old tiny chestnut horse. It could barely chew its food and a sling wrapped under its body and tied to the rafters kept the old horse on his feet. Out on the streets were troughs with wooden paddles for washing clothes and in front of the houses were large dug out squares filled with dark brown manure and water to be used as fertilizer on the fields.

The GI's slept on the floors, while the Frenchwomen kept to their own rooms. The women seemed to be living on warmed milk with bread. At chow time, Hank would eat and line up again for seconds which he would give to the woman and her daughters. On their last day at this house, Lt. Pochinchuk, Phillips, Balsam and Strecker decided to scrounge for eggs and have a feast. They found twelve eggs which the woman fried and put on a platter for them to eat at her kitchen table. They each had about four eggs.

Then there was Jeanne Thiebaut, about sixteen and blonde, with her teeth getting terribly black on the sides from poor care. She liked Hank. When they told her they were leaving the next day, she cried. In the morning when they pulled out she started to cry again. Jeanne and Hank exchanged addresses and she wrote to him about a month later. The letter had three kisses marked on it where she had sealed it and was apparently written with the aid of an army hand book of English. She asked about the fellows and hoped they would all come back safe.

* * * * *

On March 30th the outfit was trucked across the Rhine at Worms, over a pontoon bridge. The race into central Germany was on and towns fell swiftly in the rush.

In one small German town, the citizens were asked to leave as the GI's entered their homes one evening. Hank eased himself into a kitchen chair and proceeded to clean his rifle and ammo. There were around the house, quaint photos of a man in a WWI German uniform. Before Hank knew what was happening, a woman rushed into the house crying and begging Hank not to shoot her "Mann." Hank had not seen the man and tried to calm her and assure her that he had no intention of shooting her husband.

Later he had a drink of cognac, which made him sick and dizzy, so he crawled into bed and retched into his helmet—the nearest container available. When he woke up he felt better and tried to clean his helmet. There was a white ring left in it.

In one little town, the Germans were withdrawing and in the afternoon the GI's sent a patrol into the woods to observe across a river. About an hour out, they met German rifle fire. The patrol, with Sgt. Joe Juartez in charge, had met a patrol of Germans head on in a firebreak. The startled Germans yelled, "Kommen Sie hier" and Joe snapped back, "Du kommen Sie hier." Joe was fast on the trigger and shot two Germans. The others ran and the GI's returned to the main unit. They spent the night in houses in the town. Lil and Ellen had sent Hank a package of dry chicken noodle soup which he fixed on the stove in a house. It really tasted good after a steady diet of rations.

Germans slept in one part of their house while the GI's slept on the floor. Hank was having a hard time hearing and was afraid that the enemy might sneak back in on them while he couldn't hear. By morning he couldn't hear a thing and tried to scrub his ears out the best he could. That solved the problem. Orders to move came and the men left at 10:00, crossing the river and taking position in the corner of a woods where they could see the Germans on the ridge above them moving around in the bushes. An artillery officer came up and called in some rounds before the men dug in for the night.

The next day the company attacked again, moving forward about one fourth of a mile. Support was called in and one or two light tanks came up and forced the Germans out. The men moved forward, and the tanks left. A wounded German was picked up as they headed along another line of woods and stopped at a clearing.

This account was written in a notebook by Henry:

"On April 12, 1945, it was a beautiful spring day. We were fighting the Germans all day. Late in the afternoon, I was bringing up the rear of C Company."

Someone was talking and moving in the woods behind him, and Hank looked, out of curiosity, to see who was coming. One of the Germans had sneaked up on him. As he started out, he heard four or five shots. The Germans had prowled back and were sniping — and the Germans ahead of them were slowing them up too.

"The company stopped for reconnaissance, and I kept watching to the rear. We had surrounded some Germans, so I was on the look-out for them. Not to my surprise I see seven Germans — plus a few more back farther in the woods. I suppose the proper thing to do would have been to ask them to surrender but I did not. I was carrying an M-1 rifle .30 cal., with an eight shot clip. I put my rifle to my shoulder and started firing. I shot at the first one, then the second, and then the third who was carrying a machine gun.

"By that time, the rest started running into the woods. I fired the last five rounds in their direction. I then ran a few yards and re-loaded. I don't know why I made this idiotic decision, but I decided I would go back to where I shot at those Germans and if they weren't dead, I would kill them.

"I crept back and saw a helmet lying on the ground. I was going to fire at the helmet, but I thought, no, his head isn't in the helmet. So, I stepped forward a few

more steps, I saw his shoulder and chest. I raised my rifle and fired and at the same instant, one of the Germans who came to check the wounded cut loose a burst from his burp gun, but it went over my head. I ran toward the road again. At the time, I was platoon sergeant of first platoon, so I went to the head of the company to see what the holdup was. I hear about five or six shots from the rear of the company, so I go back to see what happened.

"Several of the fellows said they shot a German who was lying on the ground at our rear. I had to go check him out. I told one of the fellows to cover me while I went through his pockets. He had a picture of his wife and three children. There was no remorse from the old veteran Sgt. Strecker. I pulled up his jacket sleeves. He had two wrist watches on one arm. The fellow with me said he thought he saw a German on the path in the woods. I just said, "keep your eyes open." In the meantime, the sweat was slipping off my eyebrows. I was having a problem with the wrist-watch bands. I imagined I could feel their bullets zipping through me. Would I quit. NO. I finally got the watch loose and we joined the Company.

"It was late afternoon, so we got orders to dig in for the night. I had my foxhole dug and covered. Suddenly, I heard an explosion down in the woods. It sounded like some of our tanks were coming up along the edge of the woods. In about ten minutes, the company runner came to me and said, 'Strecker, the tank destroyer commander called our captain, and he wants us to send a squad of men into the woods to get those Germans out of there. The lieutenant wants you to take the squad.'

"Believe me, I was scared, but I got the men together and we headed for trouble. I was on the edge of the woods with one of my men and we saw two men in the woods about one hundred and fifty feet from us. We couldn't tell whether they were Germans or our men. Suddenly a shot rang out and the bullet roared like a hornet past my head.

"I just yelled, 'Let 'em have it!' The boys opened fire." Henry aimed and fired but was startled when a spray of burp gun fire flew over his head. "I jumped into the woods and got down behind a tree. The trees were young trees, about three or four inches in diameter. Once again, I emptied my rifle and once again I imagined the bullets ripping through my shoulder.

"The GI's raced into the woods where there was more cover, as the Germans behind the road bank jumped up and ran along the edge of the woods.

"The kid next to me was lying behind one of these trees and he had his leg pulled up in a V and a bullet went through the thick part of his leg and then through the calf of his leg. I jumped up and ran along the edge of the woods and they opened up on me. There was a dead branch sticking up and as I approached it, it felt like I ran my leg into the branch. The bullets were breezing *around me like bees. I knew in an instant I had been hit. The company aide man gave me a shot of morphine."*

Hank was thirsty, and his concerned buddies gave him some water. Morphine and water don't mix and soon he got sick. His squad was really upset, and the guys kept saying things like, "How will we make it without you! You can't leave us!" They were in a panic at the thought of not having an experienced man up front. He tried to reassure them. Mortars pounded them. Hank felt terrible. Here he was in a front line hole ready to get out of it—if he lived to get out—at dusk. They moved back to the firebreaks and were sitting in the woods.

Hank was half asleep and frozen when he saw what he thought was a German silhouetted on the skyline above him. Luckily, it was only another GI. Jeeps rolled up that night. Hank climbed into one and they drove back along the firebreak where the Germans were still entrenched. They reached the battalion aid station in town. An aide lieutenant asked them what the 12th Infntry Regiment boys were doing to the Krauts, since the wounded Germans he was treating were shot up so bad. Later Henry wrote, "That was my last day of combat."

The next day Hank was put in an ambulance and taken to an airfield. The wounded were being flown to Rheims. Hank had to wait while some Germans were loaded ahead of the GI's. One plane took off and Hank was put on the next. He had waited all his life for his dream to fly in a plane and now he could not enjoy it. On the way to the hospital, his bandage slipped and the rough wool army blanket grated against his open wound. He didn't care. In fact, he hoped it would get infected so that he could stay away from the war longer, or at least get to England.

At the hospital in Rheims, he was processed and sent by hospital train to Nancy. The worst off were sent to England. Henry was moved to a second-floor ward with six other men since it would take two or three weeks for his wound to heal sufficiently.

In the evening, the nurses changed the dressings and gave out clean pajamas. Hank had a bandage on his leg above the knee and a tube fixed in his wound to drain it. Vaseline was kept on it, and he received multiple penicillin shots in three hours. They ran out of saline for the penicillin, so he received the shots with distilled water instead. His arm felt like a pin cushion.

After another round of morphine, he was taken to the operating room for an anesthetic in his leg. He anxiously watched a man who had his fingers cased in wire rolled onto the gurney next to him and that made Hank feel like passing out. He could feel every stitch the doctor made into the wound above his knee. The doctor told him to stop breathing so hard or the wound would get dirty. Hank asked for a damp towel on his forehead and that helped him get through the procedure.

Back on the ward, the man in one bed next to him was a Frenchman who had been a prisoner of war working in a coal mine. He had been hit with a .20 mm. bullet and his leg was cut open from his hip to his knee. He was as pale as a ghost. A fellow in another bed was shot through the leg. Then there was "Goldbrick" called so by everyone on the ward since they all believed he was faking the back pains he professed to be suffering.

A Spanish fellow who had a cane to help him walk with his fractured shin bone would hobble up to Hank and introduce himself as "the doctor." Then he'd pretend to check Henry and would swing the cane wildly in the air, pretending to beat his leg and say, "Does that feel better?" He was good for a laugh. A fellow with both his legs in traction was staring at Hank's foot and finally asked him about the lump in his foot.

With all seriousness, Hank replied that it was a bullet the doctors had left in his foot since it wouldn't do any harm. He had the man fooled for a while with the story but finally admitted it was just a protruding vein.

At nine, Hank was given sleeping tablets and at twelve, male nurses came around and gave shots. Hank would sit up and pull his pajama top up for the shot before going back to sleep. He always felt drowsy in the morning. Hank's leg started to heal and several days later the doctor came with his scissors to snip out the stitches. It was done too soon, and the scar opened up wide.

V-E Day dawned with Hank still in the hospital recuperating. The French set off dynamite charges in the rear of the hospital to celebrate May 8, 1945. Hank was grateful but everyone was still thinking about the war with Japan.

A nurse on the ward gave Hank the dark blue Presidential Citation ribbon her nursing unit had received for their service. She said she wanted it to be worn by someone who would remember them. Henry wore it on his cap well into his eighties and proudly told people where it came from.

A young soldier with his leg off above the knee came to the ward. He would curse at anyone who tried to talk to him. A sweet old Frenchwoman was cleaning the floors one day and started to make sympathetic gestures toward him. He cursed her out and she finally had to run out of the ward. He wouldn't write home, so the Red Cross girls wrote letters for him. Hank had been lax too and wrote home to say he was OK.

* * * * *

Back in the States, his family had come home from work and school to find the neighbors telling them there was a telegram for them at the drugstore. Hit by the dread of bad news, no one would go get it. Finally, Edward, who was getting ready to go into the Army, went on his bicycle to get the telegram. He was too frightened to open it and brought it home.

It stated tersely that Henry had been wounded in Germany and other notices would follow. It was a small measure of relief at a time when telegrams could contain the worst news. The family kept all the telegrams and Henry's wartime letters.

On recovery, Hank was sent to a large replacement depot at Worms. There he developed a fever with sore throat and dizzy spells. The GI's slept in forty-man tents. Hank was sent to the infirmary which had twelve-man tents with bunk beds. After two or three days and some doses of large white pills, he was sent back to the larger tents.

He got upset when the sergeants were put on KP and ordered to clean stoves. He pulled guard duty at a brewery where the guards slept in an old boxcar at night. Finally, he was sent back to his company. As he left the depot, the major in charge was being awarded the Bronze Star for efficiency in running the camp. The GI's had a good laugh at that as they passed the parade grounds on their way out. Hank was still feeling sick and was sent to a small brick school barracks in Nuremberg for a night and the next day. They were taken to the former Nazi sports field to clean the grounds. The huge field, the cement columns, and podium were very impressive, even in their emptiness. Henry was there when the huge Nazi eagle was blown from the stadium. It was all history now.

Hank was sent to a small depot near a woods outside Nuremberg and then south to Company C in Zuggenheim for a few days. The outfit assembled in Bamberg. Hank was happy to be with his buddies again, but garrison life was creeping back. The lieutenant wanted the sergeants to sew their stripes on and the men were scolded for wearing their wool knit caps and for not being in proper uniform. There was a lot of news to catch up on. Hank had left the outfit on the road to Crailscheim and Rothenburg. Fred Jackson told Hank about how he had captured two German soldiers and some civilians in the basement of a German house after a fight right there. The two soldiers had come out of the basement with their hands up high and kept repeating—as if to justify themselves—"Vee ver shleeping… Vee ver shleeping…"

In Bamberg, there was a lot of good-natured kidding going on among the troops. Ken Thompson of Texas liked to mimic a popular radio program with, "Agh, agh, agh… the Shadow knows…" to which Hank would reply, "Yeah and the Skull's gonna take care of the Shadow. When the Skull's eyes fill with blood the Skull will strike."

Ken was in bed one day with a stye on his eye and Hank decided to act. He put on an SS skull ring, got out his SS skull pipe, (the wooden pipe was a gift from the guys in his outfit, a whole supply of which they had found in a German army warehouse) and put on a tanker's canvas hat with a skull drawn on it. He put some blood of his own in the ring's eye sockets and went into the tent, grabbed Ken by the neck and shook him yelling, "The Skull strikes now."

They were both laughing and Ken was yelling back, "Strecker, stop it, stop it." Their own "radio show" continued to be remembered in Christmas cards years later.

* * * * *

Hank was on his own until he was put in charge of a squad. Parks had taken his place. When Hank would walk past his tent, Parks would call out, "Hey, Strecker" to get him to stop and Hank would say, "Yeah, what is it?" At which Parks would answer nonchalantly, "Just checking up." Hank would start to cuss and holler just to go along with the joke, because he knew what was coming every time.

More shots were scheduled, so Hank fell into line. When he got up to the doctor and his arm was being rubbed with alcohol, he told the doctor he had already had the shot. The doctor checked his record and let him go after an officer approved. His buddies asked if he got the shots and he replied, "You don't see any pinholes in my arm, do you?"

The 12th Infantry Regiment had to practice for a parade on an air strip. At the actual parade, the three regiments of the Division were standing in rigid, neat ranks as reconnaissance planes roared overhead. Nearby, a band whistle blew, and a bass drum thundered. One soldier was so shocked by the drum that he jumped about a foot into the air and another fellow fainted in the heat. With "Eyes right" at the reviewing stand and the long lines keeping in step, Hank felt that the parade was pretty impressive.

Boxcars took them to LeHavre the next day. The weather was beautiful, and Hank watched the countryside as they went forty-and-eighting to the harbor.

At Camp Old Gold, they stayed in large perambur (pyramidal) tents. When Hank was hit, he had given his pistol to a fellow to take care of for him. He never expected to see it again. At Zuggenheim, all pistols had to be turned in. At LeHavre, a large trunk of side arms was opened, and the men were given back what they had turned in. Hank was happy to find his own pistol with its shrapnel-chipped hand grip among them.

On July 3, at 9:00 or 10:00AM, the men boarded the Sea Bass and at 5:00 in the evening pulled out. Hank watched the shores of England disappear in the dusk. That was the last of foreign lands for him. The trip home lasted seven long days. The men could hardly wait to get Stateside. Two days from New York harbor, a storm swept the seas and made the homeward bound soldiers sick. Hank stood on deck watching a glorious, yellow-cast sunset streaked with grey clouds. The water was black. Another man was sitting at the back of the ship and the reflection off the water made him look green. Hank thought he was probably sick too. (According to David Miles, this may have been his father, Ralph Jacob Miles Sr, who remembered this exact scene and also left an extensive, historical record of his own service.)

The ship arrived in a New York harbor fog. Only horns could be heard, and the buoys were the only things close enough for the anxious dogfaces to see. When the sun burned away the fog, most of the men rushed to the side of the Sea Bass where a ship carrying some USO girls was passing. Hank went to the port side. The ship was listing, and besides he wanted to see a different lady. He stared at the Statue of Liberty until it was out of sight. He thought he might never see it again. They sailed up the Hudson and under the Brooklyn Bridge. From the ship, they boarded trucks as a band played.

On the way to Camp Shanks, Hank was in the last two and a half ton truck with twelve other men. They asked the driver to step it up. He did. Hank looked ahead and noticed the line on the road swerve away in an odd direction as the truck hit a concrete abutment near a fifteen foot drop off. Everyone was piled up in the back but only one man was slightly hurt—he had a cut near his eye from broken sunglasses. The men jumped out. They had lived through the war only to have this life-threatening accident at home. The left front wheel of the truck was sheared off—luckily the back dual wheels had hit the concrete abutment, stopping the truck from going over the embankment.

The convoy commander yelled at the driver about the close call and another truck had to be sent out. At the Camp, the men were given a feast of steak, all the milk they could drink, peas, gravy, mashed potatoes, and ice cream. Hank thought it was wonderfully unbelievable real food after all the K-rations.

That night in the barracks, there was a de-briefing, and some entertainers gave an outdoor show. The men were restless and began to leave in the middle of the show until an officer got up and yelled at them to stay seated or get extra duty. Some had already left.

Hank couldn't wait to get home and took a train to Camp Atterbury in Indiana and from there to Cincinnati. From the Union Terminal, Hank hurried to Mabley & Carew, the department store in the Carew Tower where his sister Ellen worked. He asked a saleswoman to call his sister for him but not to tell her who was there. Her supervisor told her to come to the office. Ellen was worried that something was wrong or that she might have made a mistake at work. Henry waited impatiently in a small office. When Ellen walked in, Henry stood up and she burst into tears from relief, shock, and excitement.

Her supervisor let her borrow a hat, which was de-rigueur for going to a restaurant at that time. They went out for a drink and then home in a cab. At age eighty-one, Ellen could still remember vividly her excitement at seeing Henry home again.

Henry was getting ready after a thirty day leave to attend officer training school, when the war in the Pacific ended.

In late September 1945, Hank went to visit his father at the Carlton Machine Company at the request of the employees who wanted to meet a veteran, whose work they had helped supply with their machining. Henry Strecker Sr. went outside to take a photo of his son who had a big smile on his face, just like a kid. He had ridden down there on his 1932 Harley "bucket of bolts", which his brother Edward had kept in running order for him, and he was wearing his army "pink shirt" with a tanker canvas hat, his hand made leather kidney belt, tan army slacks and black motorcycle boots (which his daughter still proudly displays.) Afterward, Hank rode home and then left for discharge at Camp Butner. He never mentioned his discharge time, except on a cassette tape, describing family photos. He kept his quartermaster laundry list dated September 5, 1945 for two shirts, two handkerchiefs, two trousers, one face towel, and one barracks bag. He also kept two passes from Camp Butner to Durham, one dated October 1st to the 2nd, and one dated October 3rd to the 4th 1945.

* * * * *

When Henry retired, his daughter gave him note paper with a leather binder hoping he would write his recollections. He just could not write down all the memories which had consumed his thoughts and family conversations for so many years. It kept him awake at night. In his papers, she did find several brief notes.

"Ricochet in thigh in Normandy at Mortain.

"Concussion from potato masher grenade two feet from head in hedgerow country in July, 1944.

"Shrapnel against BAR magazine on approach to Siegfried Line. Buddy Vince Pilla killed at same time.

"Shrapnel against pistol butt in Huertgen Forest. Concussion in Huertgen Forest from shell burst three feet over hole.

"Wounded April 12th, 1945 in Germany, three weeks before war ended.

"Shave every day — think of apple blossom time in Normandy every day. Someday I'm going to quit shaving."

Photographs

Henry and Dee on his 50th high school anniversary

Wedding day, July 24, 1947

*L to R– Brothers Edward , Alan and Henry on his
graduation day 1942, Mt Healthy High School*

Ward and Dora Means Broadus, Montana

Reverend Richard Engle with Henry Strecker

Sketch from Paris by Dessin Bornett on leave December 14, 1944

Henry after the war with his old hunting rifle

Henry's grandparents from Germany, Gustav
Strecker and Elisabeth Sieckmeier

Henry's parents, Henry Strecker Sr. and Lillie Dreier

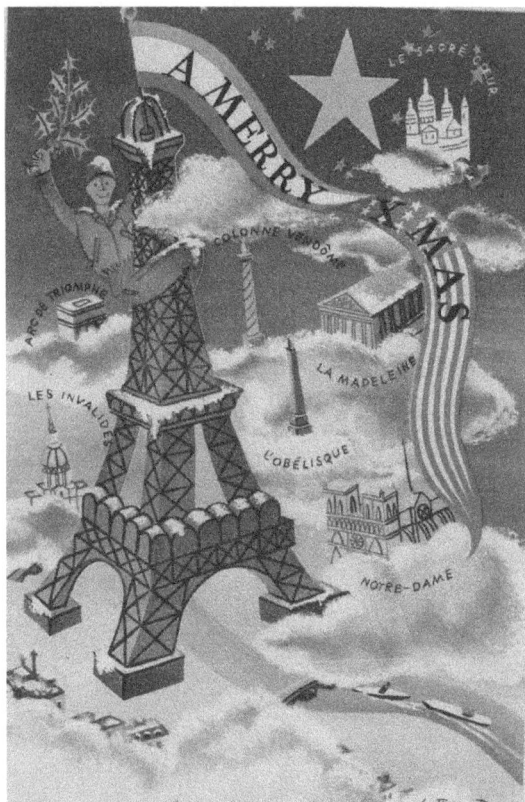

Christmas card from Paris bought on leave

Sister Ellen Strecker with Penny (buyer at Mabley and Carew)

Sister Lillian with the cat

Post Script

(added by his daughter)

After the war, Henry went to work as a truck driver at Darling and Co. which rendered dead animals and livestock for re-use, the recycling of that day. He would drive out and pick up cows, who often died under trees when hit by lightning. A favorite story of his was about being called out to pick up two horses. When he got there, he was surprised to walk up to a farmer with his two draft horses standing next to him. The farmer held out a pistol and told Henry to shoot the horses and haul them away. They were being replaced by a tractor.

It was not uncommon for farmers to get rid of anything that was no longer productive on a farm. It jolted Henry and he told the farmer, "Did the horses work hard for you many years?" and the man said, "Yes, they did." Then Henry said, "Well I cannot shoot them for you because I just got back from the war and saw way too much death and destruction — maybe you could let them remain in the pasture you have to live out their lives — because I don't want to shoot anything, ever again." The farmer took his advice.

Even years later, when mowing the lawn, Henry would slow down to give bees on the clover time to fly away.

Henry went on to other work and spent his years until age seventy as a top salesman at Aufdemkampe Hardware with the largest local accounts in Cincinnati like Milacron, Spring Grove Cemetery, and King's Island Park. Hank, salesman #5, was well known for his extensive DIY repair advice, recommendations, and efficiency, and kept a list of his customer's names in his pocket. The boss told him he could come back to work whenever he wanted. (Just missing that "Greatest Generation.)

In the mid-nineteen seventies, Hank went to the hospital for a hernia operation, accompanied by his wife Dee and daughter Leslie. Near his room, Leslie heard a nurse across the hallway loudly address a "Mr. Staubach" and she raced over to find THE Mr. Ernie Staubach, scout master of Henry's boyhood adventures for Troop 390. They shared a long overdue visit and Hank got to thank Ernie sincerely for helping him survive during the war with so many skills he learned in scouting.

More correspondence over the years followed with much emotion and healing in the unpredictable synchronicity of the universe.

Hank's brother Ed had read about an MIA soldier of the Fourth Infantry Division in a local newspaper and told him about it, thinking he might be of some help. Henry called Clark Frazier of Rawson (Findlay) Ohio near Toledo, who was searching for his missing brother. Clark drove to visit Henry and they had a heartfelt meeting over the maps and information Clark had gathered. He had gone to Germany twice, trying to find any evidence of his brother, Sgt. Billy Frazier, lost near Schlausenbach during the fighting of Sept. 19th, 1944. Clark had also assembled a large volume of military records in hopes of finding any information about Billy. He left Henry with a booklet copy of the official records he had collected on his trips to Washington D.C.

Hank sincerely told him not to keep looking because of the danger to himself of unexploded ordinance. It was an emotional day for both men, as they compared notes and maps. Clark was relentless in his search. He brought a map from the Military Archives in Suitland, Maryland along with records he was given permission to copy. Henry continued, "His brother was in my squad. He had a hard time believing me. He asked me to draw a map of foxholes we occupied up on the Siegfried Line. When he placed the map over my drawing, the foxholes lined up with his picture perfect. He didn't doubt anything more I said. He knew then that I was right there." It was an unforgettable day for both men.

Henry had seen Billy Frazier dead from a hail of shrapnel and had reported that on a list of missing in action men which they received at the end of the war in Germany. Clark made another trip back to Germany with his metal detector. He was digging with a shovel and hit the tip of an unexploded shell. Luckily it did not detonate, as many others have over the years. Clark was able to contribute to a history of local veterans "Heroes by Necessity" printed by the Hancock Historical Museum with his research and also to bury a dog tag printed with Billy's name and engraved "We have not forgotten." near a spot where he felt sure his brother had died on September 19, 1944. Henry had written about this date.

"September 19, 1944. Seven hour shoot out. Six of us left—actually four. Two wanted to surrender to the Germans but we would not let them. Our men up in the forward foxholes put their heads down in their holes and the Germans shot them while they were lying there. Standing up, I was firing a Browning Automatic rifle. I weighed about one hundred and twenty lbs. but when you're fighting for your life, you take on the strength of Goliath. About 3:00 in the afternoon the Germans gave up and withdrew. The four of us went up to the forward foxholes and our boys were all laying there dead in their foxholes. After that the Germans lambasted us with a tremendous artillery barrage. I prayed to God, twenty-four hours a day, begging Him not to let them hit me."

Sergeant Billy Frazier had sent his last letter home dated September 8, 1944. The last line read, "Had a dream last night that I was home, and Mom was baking mince pie. I'm anxious to get home and down to business. Maybe soon—Love, Billy."

Sergeant Billy Frazier is memorialized with his name inscribed at the Henrie Chappelle Cemetery; also with a

Facebook page and a book from the Hancock County Museum authored by Paulette Weiser and Ron Ammons.

Copied by Clark K. Frazier. Brother of Sgt. Billy Frazier, MIA, 19 Sept. 1944 Near Schlausenbach, Germany

C. Co. Source National Military Archives, Suitland, Maryland

File: 4th Inf. Div. 304-INF (12)7-0.3

Copied Mar. 3, 1988 & Mar. 2, 1989

CFK; 5372 CR 26, Rawson, Ohio 45881

History of Operations First Battalion 12th Infantry Regiment Fourth Infantry Division

19 June 1944 — 2 January 1945

Clark himself had a Master's degree from the University of Toledo, taught at Findlay high school, had been a farmer and was a National Champion with the National Muzzle Loading Rifle Association and in their Hall of Fame. He also was a designer and manufacturer for Matchmate Competition Rifles with one rifle at the Smithsonian and another at the National Firearms Museum. Clark passed away December 28, 2017.

* * * * *

Henry had received many Christmas cards and letters from fellow veterans after the war, until working life and building families took everyone down their own road. He saved all his post-war cards and correspondence with his old buddies. One of the most interesting was a friend who he met in the 113th Infantry F Company, Charles Reading, who continued to write to Hank from the 35th Special Forces Company and during the war overseas from the HQ Service Troop, 86th Cavalry Recon, Squad M, which was a unit assigned as entertainment. They would get military training in the morning and have practice for shows in the afternoon, learning how to work with radio, comedy, stage, and costumes. His initial unit played a large war bond concert in D.C. near the Washington Monument in September, 1943. This unit also included Baron Elliott who became a radio and live show orchestra leader for many years near Pittsburgh, Joey Adams, a comedian and writer, also on the radio, and Herb Shriner who worked in radio. Charles Reading continued to mail postcards and letters to Hank, along with a photo of the band set up in an open field at Carentan France. One note he wrote contained this clipping, "Good Dope—Your plans may all seem to go wrong, And your life may seem bitter, But son just sing the winner's song, The Lord sure hates a quitter."

Charles would pepper his letters with stage jokes that reflected his time overseas. From Germany, April 23, 1945, he wrote, "Mother Nature has taken a turn for the worse. It's been cold, rainy, sleet, and even snow. What funny April weather Germany has, the country with so many white flags. Ever since we reached here everyone in the Company is a camera fiend. I never saw so many different kinds of cameras before in my life. I suppose with all these pretty girls and the non-fertilization still in effect, the boys have to have some kind of a hobby. One of the biggest gripes at this time is the fact about us guys playing the C-ration circuit. I heard that the mess sergeants and cooks are going back to the States. The KP's will do all the work—that is, hand out the C-Ration cans. Three times a day, I wash my mess gear, but haven't eaten from it in three weeks. Our early morning show, I tried to do one of my imitations, I got a frog in my throat, truthfully speaking it was the first time in three weeks, I had any meat in my throat."

After keeping in touch throughout the war, Charles wrote from Woodside LI, NY December 31, 1948, "I had some rough luck this past year, but coming along fine at the present. I was laid up for five months with spinal meningitis. The doctors at L Hill Hospital told me I was very lucky to pull through. My sister-in-law was working as a nurse in the hospital. The nurses all took good care of me, and treated me as one of the family. It has been a mean year for me. Looking forward to a bright future. I'm building up a new act with a friend of mine in show business who sings. I'm working a night spot tonight "New Years Eve" sort of breaking it in for the first time. It's a clean act. Thanks a million for keeping up our friendship since 1942."

My father's last card from him was postmarked Flushing, New York, dated January 7, 1950 from Woodside, Long Island. He wrote, "The only reason I'm late getting this note to you, I was working in Saratoga for the holidays, in a place called Newman's Lake House, just returned a few days ago. I'm kind of late saying Congratulations on your ten- month-old baby girl. Good luck and success in your home life … May this be the happiest one you have ever had. If you get to New York, don't forget my address and be sure to look me up. Your pal, Charlie." Sadly, Hank never heard from Charles again.

He did continue to receive letters and Christmas cards from many of the men he had served with including: Edmond Phillips, Paul Hackley, Robert Maher, Robert Grey, Fred Jackson, and Ken Thompson.

In 1970, Henry and Dee worked at the Cincinnati, Ohio reunion of the Fourth Infantry Division Association—their first reunion. It was exciting and exhausting. After their daughter graduated from college, she sent them to the Buffalo, New York reunion where they had an amazing, wonderful time and got to enjoy Niagara Falls. As they went to more reunions over the years, they gained close friends. At Christmas, Dee would send over a hundred cards to the veterans families that they had met. They also met World War I and Vietnam veterans and Henry, along with Dee, would always welcome the new members. Once she found an older veteran alone in a reunion hotel lobby looking miserable, so she escorted him to the hospitality room and he told her later, he had the best time of his life meeting the other veterans. Later he sent Henry and Dee a huge box of Christmas nuts and candy as a thank you.

At another convention, one of the first Vietnam veterans to attend was warmly welcomed by Henry. They became very good friends. Vern Capoeman was a Quinault Indian and had come all the way from his home in Washington state. He continued to write to Henry over the years and also sent him several beautiful feathers, including an eagle feather, which he was permitted to do. It was proudly framed. Vern passed away, April 13, 2013, age sixty-six.

One of the best reunions for Henry was meeting Captain Sullivan and his wife "Tiny." The Captain had believed Henry was killed in a shell burst in the Huertgen Forest and was shocked when he saw Henry standing there at a reunion. They stayed in touch over the years and Dee remained friends with many veterans' wives and enjoyed their friendship at so many yearly reunions. Henry's daughter also stays in touch with some of her father's buddies' children and grandchildren, as do many other veterans' families by mail and through The National 4th Infantry Division Association on line, maintained by Sue and Bruce Gass. It is kind of like having third cousins from a foxhole.

"The Ivy Leaves" magazine was a much-anticipated mailbox arrival from the National Fourth Infantry Division Association and Henry would pour over it at the kitchen table and saved copies from 1954 onward. Finally, his daughter sent some of his memories in to Rusty Armstrong, the editor and a 4ID Vietnam veteran. The March 2003 surprise issue arrived, and Hank turned the pages to read with disbelief, "by Henry C. Strecker."

He took a deep breath of overwhelmed appreciation.

* * * * *

In his own words, more memories were recorded:

"Over the years, especially in November when the snow fell, I would remember Brownie Means, the man from Big Sky country Montana and often wondered if he was ever able to ride horses again, which was what he talked about and missed the most during the war. It was sad not knowing if he even lived through the war.

"In 1992, I attended my fiftieth Mt. Healthy High School reunion. During the celebration dinner, people got up and told a little about themselves. There were only seventy-eight graduates in the 1942 high school class and one woman got up and mentioned that she was from Broadus, Montana.

"I remembered that by a strange coincidence that was where Brownie Means had lived and later asked if she knew him or his family. She said she knew a Ward Means and the next time she was home in Broadus, she would try to give him my phone number and address. A few months later, she did see Ward in the local hardware store in Broadus. He had moved to Sheridan, Wyoming, but was visiting Broadus. Ward called me, and he and his wife drove down to visit us in Ohio. It was a very emotional visit after almost fifty years."

Ward and Dora Means stayed in touch with Henry and Dee for the rest of their lives. Ward offered to put Hank on a wild horse. Although he had terrible back pain all his life, Ward was able to work with horses while Dora raised the children.

The spirit of "Steadfast and Loyal" lives on.

Henry continued with special memories:

"On May 25th, 1994, my wife Dee and I went to a mini-reunion of the Fourth Infantry Division at the Radisson Hotel in Columbus, Ohio. It was a four day get together and there were several other conventions going on in the same hotel.

On Friday evening we were in the hospitality room on the second floor. Dee and I always like to welcome new members. I had just met a first-time attendee and while we were talking, a man with a gray beard and wearing a business suit walked in and then left. After I finished talking to the man I had just met, I asked the bartender if he knew the name of the other man who had just left the room. He said he thought it was Angle or something like that. A light went on in my head as I said, "Was his first name Richard?" and he said he didn't know. I ran straight down to the elevator and entered the lobby. One of the men who organized the reunion was walking toward me and I asked him if he had seen a man with a beard. He said yes, and he had tried to get him to join the Fourth Infantry Division

Association. I asked what his name was, and he said, "Richard Engle." I was frantic. I thought after fifty years of wondering if this man had lived, he got away from me.

"I went to the girl at the desk and asked if she had a listing for Richard Engle. She said yes, but I can't give you his room or phone number. I explained my situation and she said, she would try to call his room. After about an hour, I gave up. I figured he had gone somewhere, and I would check again at about nine the next morning. That morning the operator connected me to his room.

"My heart was pounding. I said, 'Richard, you won't believe who this is.' After I explained our last encounter in the hedgerows, he said he would come right down to our room. We had a wonderful conversation for about an hour. They had sent him right back home after he got hit. The doctors told him if he would have coughed on the way back from the front, he would have bled to death. Needless to say, we were two grateful buddies. Richard had become a priest in the Columbus, Ohio area and in addition to being a pastor, had also worked at VA hospitals. He was only at the Radisson that day to attend a Catholic convention.

"When he saw the Fourth Infantry Division convention sign, he had stopped in out of curiosity. He had the bullet Henry handed him and carried it with him. Richard Engle was good friends with our parish priest

and would visit him and my parents when he was in town. He credited his own Mother's prayers with saving him during the war and went by the nickname "Bird" from high school days when he notably bird watched. In the days before the internet, we were overwhelmed by their "by the grace of God" reunion."

"Father Bird" reflected on his time in the war, "My mother spent an hour every day at St. Patrick's Church in Columbus, praying for me and my brother Jack. I'm convinced her prayers saved my life." He carried the bullet casing in his pocket throughout his life and remembered that day in 1944, "One hit me in my left shoulder and came out my back. If it would have gone an inch the wrong way, it would have gone through my chest and into my heart and I'd be dead. The other one hit my helmet and my right arm. It could just as easily have gone through my brain, but it also hit the right spot."

As a side note, after Henry married Dee, he was going bowling two nights a week and she thought that was rather a lot. What she didn't know, was that on the one night he was actually studying to become a Catholic like her Italian family and most of his German family. Father Bird attended Hank and Dee's 50th wedding anniversary dinner to give a blessing and proudly showed us the shell casing that went through him in Normandy. During his introduction, he mentioned their connection to the 4th Infantry Division.

Father Bird gave Henry a final blessing when he became ill for the last time and continued his friendship with our family over the years until he passed away on April 21, 2021.

The National Fourth Infantry Division Association magazine contains an often very meaningful Letters to the Editor section. That is how my father met more veterans to correspond with over the years. Their personal stories have added so much to the Division history.

Henry's connection to Marcus Dillard began when both of their letters were published side-by-side in the Ivy Leaves dated March 2005. Marcus wrote to Rusty Armstrong:

"Dear Editor, I noticed an individual wrote in the December 2002 issue of the Ivy Leaves asking about the background color of the 4th ID patch during the War (WWII).

"Well, I have one, just one and the reverse of the patch is cross stitched and the green leaves of the patch has an olive drab background with the dark green leaves on the patch. This one patch has a sentimental value attached to it. I had met this beautiful German girl where we halted briefly during the end of the war. I think I fell in love with her the minute I saw her! Anyway, after about two weeks we received orders to leave at once. I found her real quickly and told her we had to leave. I ripped off this patch of the 4th Infantry Division and

told her to keep it to remember me by and that I would be back to get her and the patch. I DID! And we were happily married for over fifty years."

Following are excerpts from letters written to Henry Strecker from Marcus Dillard, WWII, Company M, 3rd Battalion, 12th Regiment, 4th Infantry Division. Together they remembered the same locations overseas. Marcus had also been interviewed by Peter Jennings for his account of being in Normandy and Paris. Marcus passed away July 1, 2007 and he, like so many is deeply missed. Internet search his name and many articles will appear. The following letters were written to Henry and some to his daughter, Leslie.

Letter dated March 7, 2007 …

Well, I have another little story to relate to you, that you may find hard to believe. It was during the Battle of the Bulge. We were moving up in the Ardennes in the dead of night. It was cold, cold, cold. I am sure your Dad remembers. Full moon, snow so frozen, that it went crunch, crunch every step you took. Suddenly, we saw some object over to our right in the snow. We got closer and what we saw was a German soldier and an American soldier sitting back to back. They were dead naturally, but we did not know how they died. Maybe they were wounded and got together the way we found them, to keep each other warm. Ironic in a way, enemies, but got together to try to survive … "

Letter dated September 9, 2005 ...

"I do not think you could ever put into words how terrible the Huertgen Forest was unless you were there. The morning we arrived ... raining, cold wind blowing, dark, miserable, soaking wet, not knowing where you were. I was nineteen years old." ... and from another letter dated Feb 6, 2007, "In the Huertgen, we stayed with our 81mm. mortars about 500 to 1000 yards behind the line companies to give them our fire support. ... "we had to dig our mortars in about four feet deep and then we had our own foxholes where we tried to sleep. We made it into a small bunker covered with logs and dirt. We then took our metal containers that the shells for our mortars came in, punched holes in the containers, poured gasoline into the containers after we filled them half full of dirt, put a match to it and we had a small stove, but a lot of black smoke. Our faces were always black from the smoke.

Letter dated April 1, 2007 ...

the German commander said, "It will be easy to take Dickweiler." Was he ever wrong. I was in Dickweiler with our 81 mm. mortars supporting. Dickweiler sat at the bottom of hills, just like sitting on the bottom of a tea cup. Anyway, the Germans came, they had to come down the hill toward us, totally exposed. We cut them to ribbons. We were surrounded but they could not get in. . . . It is interesting to note after sixty-three years I found out that your father (Henry Strecker, Co. C) was one of those sent to Dickweiler to help us. I have another little tale to tell you about Dickweiler. We were surrounded and after a while, things quieted down. I was in the tallest house in the village perched on a small hill and I was looking out one of the back windows toward the road that led to Herborn where our company had our CP. What I saw shocked me. I saw a lone German soldier carrying a small machine, walking along like on a Sunday stroll. I pointed him out to my buddy, Roland Carmack, who was from Alabama and a crack shot. I would say the distance was 800 or 900 yards. I had field glasses, and I could see him well. My buddy shot and the German rolled down a little bank beside the road. We could not tell if he was hurt or not. I used the glasses and told Carmack where to shoot, whether to raise or lower his elevation or what. The German laid there all day. The next morning, he was gone. We do not know if the Germans took him during the night or not... Warmest regards, Marcus."

Reading The Ivy Leaves, Henry saw a letter from Hubert Gees who had served in the German military opposite him. Henry corresponded with him yearly at Christmas. Hubert had a very compelling story, having become a POW in the Huertgenwald and was sent to America as a prisoner of war. He learned to play baseball with the camp guards and became friends with one. The American invited Hubert to visit after the war and he did, adding to better memories. Herr Gees sent us photos and maps about the Huertgenwald, to compare his position there with Henry's. Hank's daughter was very happy to translate his letters since she had studied German in college.

In 2004, Hubert's Christmas letter was so heart-felt, she could barely read it to Henry through all their tears.

December 9, 2004

Dear Henry, wife Leanora and daughter Leslie,

How are you?

Dear Henry, in the fall of 1944 we could not imagine that after sixty years, we would have a binding friendship with a soldier from the other side ... .

... We Germans have America to thank that after the end of World War II, we were taken into the circle of free nations and that we became a free democracy.

That we must never forget!

So, our hearts are with the American soldiers in Iraq, who are making the bloody sacrifice in battle with terrorists and fanatics.

And so, once again, the soldiers of the 4th US Infantry Division, for the New Year will try to bring peace to the Near East.

Merry Christmas and a peace filled year 2005.
Yours, Hubert Gees and wife Thea

Hubert shared his memories of the Huertgenwald where he served as a company runner with the 2nd Co., Fusilier Bn., 275th Infantry Division. He was a speaker at the dedication of a monument to German Lt. Lengfeld in the Huertgenwald, which was begun by Col. John F. Ruggles (retired as Major General.) Lt. Lengfeld had died trying to rescue an American who was injured there in a minefield. One of the best articles about the monument in the Huertgenwald, which was sponsored by the 22nd Infantry Regiment of the 4th Infantry Division members is by Bob Babcock, 4ID veteran, historian, author, and publisher. To read more see: 1-22infantry.org/history4/lengfeld.htm.

In 1997, Hubert Gees wrote "There are always more of the fallen dead being found in the Huertgen Forest." Many were buried under layers of trees from the artillery barrages during the battles there. He also sent maps of his location near the Weisser Weh' and sent newspaper clippings which my father was grateful for, as he drew a map of his own location to exchange with Hubert. As a side note the latest identified casualty was Staff Sgt. Max Thurston 28th Infantry, who died November 6, 1944 and was originally found in April of 1948. On July 7, 2023 his remains were identified by the Defense POW/MIA Accounting Agency. There is more information on the web at "New Stories from the Huertgen Forest" and "Current News from the Huertgen."

December 9th, 2000, Hubert wrote again about the Huertgenwald, which he had re-visited several times.

"Dear Henry, ...

In the early afternoon of November 17, 1944, on the site of today's military cemetery, we were trying to get cover from American artillery shells and jumped into a large crater in the ground. I suddenly encountered a dead American soldier. I will never forget the sight of him with his facial features still intact. With wide open eyes he stared at the sky above Huertgen. His open mouth seemed to have called out. I was not indifferent to this dead man, and I asked myself what was the point of our killing each other. "Where in wide America will a mother mourn for him?" I wondered lying right next to him.

My Christian faith gives me the hope that soon in God's eternity, I will be able to extend my hand to him for reconciliation. Christmas, the Christian festival of peace, is near and time to commemorate all those who lost their hopeful lives in the Huertgen Forest and on other battlefields. May you all rest in God's everlasting peace! ...

Yours, Hubert Gees and Frau Thea"

Hubert (who had a pacemaker for many years, passed away on September 30, 2013.

* * * * *

Through the years on Memorial Day my father would go alone to place wreaths, provided by the local Ivy Division chapter, on the graves of Ivy veterans around our city and also up near Dayton. His daughter asked to go with him for years, but he always said no, and she was resigned to his mission, his isolation, and also proud of him.

As he aged and could not drive, our family attended local Memorial and Veteran's Day events with Henry. Many were held at our local cemetery with high school bands, military speakers, retired veterans, prayers, and ceremonies. Henry would wear his WWII uniform and hat. After one occasion, an elderly couple came up and asked to shake Henry's hand. They were visiting our area and thanked him profusely for his service. To our amazement, they had been residents of Paris on August 25, 1944 and remembered with the excitement of their youth that day the 4th Infantry Division liberated Paris. The three of them once again beamed with the strength of 1944.

It seemed surreal, not a coincidence, but once again the very real Hand of Divine Providence in Hank's life.

A very memorable family trip was to Rickenbacker Field for The Gathering, an air show featuring spectacular P-51's and many vintage and top of the line military aircraft. It was September 30, 2007. Dad wore his uniform and was in a wheelchair to better navigate the huge venue. In addition to seeing so many great aircraft close-up, we saw a Raptor fly above us, hover and open its bomb bay with the smoothness of a cigarette case. Truly impressive, with all due respect, it looks like death. Then the P-51's had a roaring, unforgettable flyover. When the Thunderbirds blasted low over our heads, it took our breath away. Henry cheered with amazement, followed by ear-to-ear grins with his family. He was beyond happy.

The whole day fulfilled a lot of his life-long dreams about airplanes. Children and little ones, at the direction of their parents, kept coming up to him respectfully, to thank him for his service or politely shaking his hand. It was a very emotional day, capped off by being greeted and sincerely welcomed by the 101st Airborne World War II Reenactors, led by Robert Traphagan, who now directs the Central Ohio Military Museum. Their conversations with Henry and having him hold their old M-1 rifle tapped off a truly unforgettable day. As we left, we purchased a souvenir book and found out that one guest of honor that day was Robert J. Frisch, a WWII Mustang P-51 ace ('Worry Not' was his plane). Too late in the day to visit, my father told me he *had graduated from Mt. Healthy High School with him in 1942.*

* * * * *

The following is a note Henry wrote a year after retiring:

"I fought ten months of that war. I received three con–

*cussions that create a weird taste in your mouth from the
blast of one grenade and two different artillery shells.
The taste is indescribable. The fluids in your head are
blown through your mouth. I was hit four times, caught
a ricochet in the right thigh in Normandy, on the way
to the Siegfried Line one of my BAR magazines stopped
a shell fragment. I carried a German Sauer 7.62 pistol
on a shoulder holster that stopped another shell fragment
in the Battle of Huertgen Forest. On April 12th 1945,
they finally got me with a machine gun bullet across my
left leg above the knee. I spent the entire war outside,
rain, snow, freezing cold, and mud. I fight the battle
over every day for the past forty-four years. I can be a
trembling man in ten minutes, and I thank God for the
warm bed I lie upon at night. I'm not crazy yet, but I'll
fight that war to my grave."*

March 6, 2021, was the 11th anniversary of my father's
last day in 2010. That evening, I was reading over my father's notes about his last days in combat which were very
gruesome and troubling. Several times he had told me that
when he fired point blank at German soldiers who fell, he
would convince himself that someone else had fired the
fatal shots. Adding to this grief was the loss of so many
good men engraved in his memory. I could never picture
my kind, gentle, courteous, humor-filled father in the kind
of combat situations he often described.

His thoughts on that, written by himself are very telling

of a person trying to reason the unreasonable with himself. He refused to go to the VA, although we urged him to as he got older, to make some connections with other veterans when he could no longer attend reunions or have a counselor to confide in with his emotions.

As is so true, no one can really understand this situation unless they have been there themselves. I found a quote by Bill Mauldin, "The combat man isn't some clean-cut lad, because you don't fight a kraut by Marquess of Queensberry rules. You shoot him in the back, you blow him apart with mines, you kill or maim him the quickest and most effective way you can with the least danger to yourself. He does the same to you. He tricks you and cheats you, and if you don't beat him at his own game, you don't live to appreciate your own nobleness. But you don't become a killer. No normal man who has smelled and associated with death ever wants to see any more of it. In fact, the only men who are even going to want to bloody noses in a fist-fight after this war will be those who want people to think they were tough combat men, when they weren't. The surest way to become a pacifist is to join the infantry."

I fell asleep the night of March 6, 2021, troubled by Dad's suffering and tears. The next day I woke up and started to read the Winter 2020 issue of The Ivy Leaves and came across two letters from Vietnam veterans, Dennis Lewallen, 4ID,1st,12th, Co. D, 1967-68 and Edward Powell, 4ID, 35th Inf, Co. B., 1967-Dak To. As I read, I realized that their letters gave me answers I needed about my father's life. Each veterans' thoughts gave me real peace and understanding about my father's own suffering, separated by time from them, yet united by "Steadfast and Loyal." This is remarkable to me for many reasons. I found their letters to read on the perfect day and in perfect relationship to his memories. The Ivy Leaves had connected me to my father again.

I found another note in my father's kitchen cupboard letting me know he had reconciled with himself over his last day in combat and killing on April 12th, 1945. In a letter written to Ward Means' son, Randy Harvey, he wrote, "I never thought I would ever try to kill a man when he was lying there wounded, but I did. I would not want someone to kill me if I were lying there wounded. God said—Strecker you're not thinking straight anymore so I'm going to take you out of this nightmare. This time I AM going to let you get hit—which HE did."

Post Post Script

Dad's adventures as a boy, included finding an injured hawk he took into their basement, to help it heal. He fed it ground beef and one day it bit him so hard he bled and learned a valuable lesson. Leather gloves may not be enough protection from a hungry, stubborn hawk. He never forgot how bad that hurt but as the spring came, the hawk gained strength and Henry released it out a basement window. Hawky the hawk, flew away hail and hardy. It was hilarious to see Dad, over the years, imitate the hawk crunching down into his hand with all the power it could muster. Hawky became a family legend. Leslie would kid Henry and tell him that the progeny of Hawky were all the hawks that preyed along the new highway by his home.

When he passed away, it was an overwhelming, grief-filled time for our family. He died quietly at home, just stopped breathing in his hospice bed after eighteen months with Parkinson's, as Mom and I stood on each side of him. It was almost six in the evening of March 6, 2010. I visited Mom often, and somewhere during her move a year later,

I went to her home to gather some items for her new residence. I walked over to the front door, I knew, for one of the last times. There in the center of the steps, lay a beautiful hawk feather, as if it fell there with a purpose. I had never seen a hawk feather. Such a perfect hello and good-bye in the synchronicity of the universe. Divine Providence.

About the Author

Leslie Strecker's life was shaped by the love and hard work of her parents, Dee and Henry Strecker. They had very little in material possessions but evidenced much more happiness in daily life than many others. They dedicated themselves to participating in their children's lives, with talks around the kitchen table about history, nature, astronomy, reading classics and opening every door they could to help Leslie and son Steve grow into adults who wanted to learn every day. Leslie began to write down her father's experiences from World War II with the 4th Infantry Division, starting in sixth grade. Her parents' world is gone, but their memories and time will be remembered. D-Day will remain pivotal in the synchronicity of time, just as are the far distant memories of the Trojan Wars.